GLOBAL AGRICULTURE DEVELOPMENTS

CROP YIELDS

PRODUCTION, MANAGEMENT PRACTICES AND IMPACT OF CLIMATE CHANGE

GLOBAL AGRICULTURE DEVELOPMENTS

Additional books in this series can be found on Nova's website under the Series tab.

Additional e-books in this series can be found on Nova's website under the e-book tab.

CLIMATE CHANGE AND ITS CAUSES, EFFECTS AND PREDICTION

Additional books in this series can be found on Nova's website under the Series tab.

Additional e-books in this series can be found on Nova's website under the e-book tab.

GLOBAL AGRICULTURE DEVELOPMENTS

CROP YIELDS

PRODUCTION, MANAGEMENT PRACTICES AND IMPACT OF CLIMATE CHANGE

LIJUAN HUANG
AND
QIAO ZHAO
EDITORS

New York

For permission to use material from this book please contact us:
Telephone 631-231-7269; Fax 631-231-8175
Web Site: http://www.novapublishers.com

NOTICE TO THE READER

Library of Congress Cataloging-in-Publication Data

Crop yields : production, management practices, and impact of climate change / editors: Lijuan Huang and Qiao Zhao.
p. cm.
Includes index.
ISBN 978-1-62618-175-5 (soft cover)
1. Crops and climate. 2. Crop yields. I. Huang, Lijuan, 1952- II. Zhao, Qiao, 1954-
S600.7.C54C765 2013
338.1--dc23

2013004726

Published by Nova Science Publishers, Inc. † New York

CONTENTS

PREFACE

In this book the authors present current research in the study of the production, management practices and impact of climate change on crop yields across the globe. Topics discussed include the peculiarities of photosynthesis on wheat genotypes contrast in grain yield and their use in breeding programs; options for optimal use of water resources for securing future agricultural production; the potential impact of climate change on wheat production in Australia; and improvement of cotton production in arid saline soils by beneficial microbes.

Chapter 1 - 60 years of comprehensive research on photosynthesis and productivity of various wheat genotypes in natural conditions of cultivation, characteristics and parameters of photosynthetic activity of these genotypes in crop fields have finally established the high grain yield of an “ideal type” of wheat variety. For this purpose, the rich genefund, comprising several thousand wheat genotypes, selected from both the ancient and aboriginal varieties of national selection and introduced from the world genefund, particularly from *CIMMYT, ICARDA* and other International Centers, with contrasting grain yield, photosynthetic traits (stem height, area and architectonics of the leaf surface, etc.), duration of the vegetation, and other morphophysilogical traits, as well as drought resistance, was created. These genotypes, with two or three times less leaf areas than those with broad leaves, produce similar or even greater grain yield. Analysis of various wheat genotypes with different values of photosynthetic traits and productivity in conjunction with a range of environmental factors, including mineral nutrition, water, light, etc. showed the wide range of CO_2 assimilation variability in ontogenesis, depending on the morphophysiological peculiarities of genotypes and their sink-source relations. The diurnal changes in the photosynthetic rate of leaves of all layers and genotypes with contrast architectonics are characterized by a double-peaked pattern indicating a drastic increase in the photosynthetic rate in the morning and a decrease in the evening. Not all genotypes with small leaves are highly productive and not all genotypes with broad leaves are considered high or low productive. Genotypes with broad leaves and high yield require sufficient water supply. Genotypes with vertically oriented short and narrow leaves and a high tolerance to water stress yield up to 10 t ha^{-1}. The rate of biosynthesis of the main transport form of carbon, sucrose, and products of glycolate metabolism as well as the rate of CO_2 release in the light due to photorespiration were higher in highly productive genotypes. The lower leaves of highly productive genotypes export more assimilates to the ear than lower leaves of low productive ones. Favorable architectonics that provide optimal assimilating surface and sufficient attractive force of genotypes in sowings with high photosynthetic activity of the ear create the basis for a high grain yield. Throughout the entire period of flag leaf development, highly productive intensive genotypes in comparison with extensive ones were distinguished by higher activities of RuBP carboxylase and carbonic anhydrase that play a significant role for the maintenance of CO_2 assimilation on the high level. The activities of the RuBP oxygenase were higher in the highly productive wheat genotypes than in the lower ones. The activity of RuBP carboxylase in different ear elements varies deeply depending on the development and genotypical features of the plants. The ear glume and awns of the intensive genotype had a higher activity of RuBP carboxylase than those of the extensive genotype. Thus, the high level of metabolism, high activity of the key enzymes and the primary photochemical processes, in conjunction with favorable phenotypic traits, optimum leaf index and architectonics are crucial to the high productivity of wheat genotypes.

Chapter 2 - Agriculture is the main user of fresh water resources in the World. However, irrigation water has become less available in many areas due to an increased competition for water among industries, households and the environment and due to global climate change. Thus, there is a growing interest of using irrigation water resources in a more efficient manner. Here a viewpoint is made by looking on the efforts and knowledge already existing about optimising the use of irrigation water resources. Main tools for optimising the use of water resources in agriculture can be obtained by: selection of water efficient irrigation methods like drip and micro irrigation, use of low quality waters such as treated wastewater, use of salt tolerant crops, optimising irrigation scheduling by using plant and soil sensors, remote sensing, irrigation models, using the 'virtual water' principles so that water rich regions secure food supply to dry regions. In dry regions small areas with high value crops should be irrigated securing net income and livelihood. In general water can be saved by reducing the overall waste of food throughout the food chain from post-harvest to the end-consumer. Recent findings on these options are discussed and evaluated.

Chapter 3 - The potential impact of climate change on wheat production is a key concern to Australia due to its importance to the nation's economy and to global food security as wheat is a key player in the world grain market. This work extensively reviewed the relationship between wheat crop production including yield and future climatic change based on both experimental and modelling studies conducted in Australia over the last four decades. Experiment studies evolved from adopting glasshouse, gradient tunnel facility to free air CO_2 enrichment facility, evolved from studying enhanced CO_2 effect alone to the combined effects of increased atmospheric CO_2 concentration along with other environmental factors such as light, temperature, soil water and nutrient, and with various soil types, and evolved from wheat monoculture to wheat-pulse rotation system. Modelling approaches evolved from sensitivity analysis, to scenario analysis and to Bayesian analysis. The advantage and disadvantage of each research method were discussed where appropriate. Future research directions were identified to better address future climate change, food security and sustainability issues.

Chapter 4 - Crop cultivation in saline soils is one of the major agricultural challenges world-wide. Cotton production in Uzbekistan is limited by soil salinization and disease caused by soil borne pathogens. The use of plant growth promoting rhizobacteria (PGPR) to control plant diseases and stimulation of plant growth has been considered a viable alternative and environmentally friendly method. The strain *P. putida* BIST and *B. subtilis* SUBTIN significantly ($p<0.05$) increased the total N, P, and K contents of shoot and root (13-27%), plant height (15%), and yield (16%) of cotton in small plot experiments. They were able to produce phytohormone indole-acetic-acid (IAA) under saline conditions and were positive for ACC deaminase activity. On the basis of results, it may be concluded that plant growth promoting rhizobacteria are potential option for improvement of cotton growth and development in salinated arid soil.

In: Crop Yields
Editors: Lijuan Huang and Qiao Zhao
ISBN: 978-1-62618-175-5

Chapter 1

PECULIARITIES OF PHOTOSYNTHESIS OF WHEAT GENOTYPES CONTRAST IN GRAIN YIELD AND THEIR USE IN BREEDING PROGRAMS

Jalal A. Aliyev[1,2,*]
[1]Department of Plant Physiology and Biotechnology,
Research Institute of Crop Husbandry,
Ministry of Agriculture of Azerbaijan Republic, Baku, Azerbaijan
[2] Department of Fundamental Problems of Biological Productivity,
Institute of Botany, Azerbaijan National Academy of Sciences,
Baku, Azerbaijan

ABSTRACT

60 years of comprehensive research on photosynthesis and productivity of various wheat genotypes in natural conditions of cultivation, characteristics and parameters of photosynthetic activity of these genotypes in crop fields have finally established the high grain yield of an "ideal type" of wheat variety. For this purpose, the rich genefund, comprising several thousand wheat genotypes, selected from both the ancient and aboriginal varieties of national selection and introduced from the world genefund, particularly from *CIMMYT, ICARDA* and other

* Email: aliyev-j@botany-az.org.

International Centers, with contrasting grain yield, photosynthetic traits (stem height, area and architectonics of the leaf surface, etc.), duration of the vegetation, and other morphophysilogical traits, as well as drought resistance, was created. These genotypes, with two or three times less leaf areas than those with broad leaves, produce similar or even greater grain yield. Analysis of various wheat genotypes with different values of photosynthetic traits and productivity in conjunction with a range of environmental factors, including mineral nutrition, water, light, etc. showed the wide range of CO_2 assimilation variability in ontogenesis, depending on the morphophysiological peculiarities of genotypes and their sink-source relations.

The diurnal changes in the photosynthetic rate of leaves of all layers and genotypes with contrast architectonics are characterized by a double-peaked pattern indicating a drastic increase in the photosynthetic rate in the morning and a decrease in the evening. Not all genotypes with small leaves are highly productive and not all genotypes with broad leaves are considered high or low productive.

Genotypes with broad leaves and high yield require sufficient water supply. Genotypes with vertically oriented short and narrow leaves and a high tolerance to water stress yield up to 10 t ha^{-1}. The rate of biosynthesis of the main transport form of carbon, sucrose, and products of glycolate metabolism as well as the rate of CO_2 release in the light due to photorespiration were higher in highly productive genotypes. The lower leaves of highly productive genotypes export more assimilates to the ear than lower leaves of low productive ones.

Favorable architectonics that provide optimal assimilating surface and sufficient attractive force of genotypes in sowings with high photosynthetic activity of the ear create the basis for a high grain yield. Throughout the entire period of flag leaf development, highly productive intensive genotypes in comparison with extensive ones were distinguished by higher activities of RuBP carboxylase and carbonic anhydrase that play a significant role for the maintenance of CO_2 assimilation on the high level.

The activities of the RuBP oxygenase were higher in the highly productive wheat genotypes than in the lower ones. The activity of RuBP carboxylase in different ear elements varies deeply depending on the development and genotypical features of the plants. The ear glume and awns of the intensive genotype had a higher activity of RuBP carboxylase than those of the extensive genotype. Thus, the high level of metabolism, high activity of the key enzymes and the primary photochemical processes, in conjunction with favorable phenotypic traits, optimum leaf index and architectonics are crucial to the high productivity of wheat genotypes.

Keywords: Wheat genotypes, photosynthesis, grain yield, carbon metabolism, transport, distribution, assimilates, enzyme activity, carbonic anhydrase, phosphoenolpyruvate carboxylase, ribulose-1,5-bisphosphate carboxylase, ribulose-1,5-bisposphate oxyganese, ear, glyme, awn, protein

INTRODUCTION

Photosynthesis is the major determinant of plant productivity, and its investigation is of principal interest for further increasing crop yields (Nichiporovich, 1982; Aliev, 1974, 1983; Aliyev, 2012). The realization that crop yields are reaching a plateau, while population increases continue at pace, has placed manipulation of photosynthesis in a central position to achieve increases in yield (Raines, 2011). Increased efficiency of photosynthate production and utilization is an attractive approach to increasing crop yields. Plant productivity and crop yields are governed by the rate and duration of photosynthesis and the partitioning of fixed carbon between sucrose and starch in source leaves as well as the capacity of sink tissues to assimilate sugars and amino acids from source leaves. Many researchers investigated the relationships between plant productivity and photosynthesis by studying photosynthetic performance in crop varieties differing in their productivity. This interest explains the numerous attempts to raise crop yields by improving photosynthetic activity through plant breeding (Bykov and Zalenskii, 1982). This approach motivated investigations of energy metabolism in plants differing in productivity. In particular, it has been shown that the photochemical activity of chloroplasts from high-yielding varieties is 30-40% higher than that of chloroplasts from low-yielding varieties (Volodarskii, 1980). The activity of cyclic and non-cyclic photophosphorylation was also shown to be higher by a factor of 1.5-2 in high-yielding varieties (Kazibekova et al., 1985).

It should be emphasized that a mere increase in the photosynthetic rate does not necessarily lead to higher grain production, because the economic yield is determined not only by the CO_2 assimilation rate, but also by the ability of plants to export photosynthates from leaves and store them in economically valuable parts of the plant (Kursanov, 1984). Moreover, the transport and distribution of assimilates are closely linked to photosynthesis and actively influence the photosynthetic rate (Mokronosov, 1981). In turn, the transport and distribution of assimilates depend on sink source relations between organs of the whole plant. Although the patterns of assimilate distribution is genetically determined, it markedly changes upon natural or artificial alterations in sink source relations. Therefore, the current trend in physiological research is to find the means to regulate the transport and distribution of assimilates by studying these processes in diverse varieties and cultivars (Kursanov, 1984, Bonnemain, 1982).

Though in some papers a direct correlation between photosynthetic indices and plant productivity has been neglected (Evans and Dunstone, 1970, Milthorp, 1978), the comparative investigation of photosynthetic activity of cultivars differing in productivity showed that the new highly productive cultivars had a more intensive photosynthesis (Nichiporovich, 1956, Aliev, 1974, Kumakov, 1977, Aliev and Kazibekova, 1979 a, b, Bykov and Zelenskii, 1987), higher rates of electron transport and of cyclic and non-cyclic phosphorylation, as well as more powerful regeneration of reduction potential (Gavrilenko et al., 1974, Volodarskii et al., 1980, Kazibekova et al., 1985). Unfortunately, the investigation of photosynthesis in relation to production was basically carried out at the level of primary processes of radiant energy absorption and transformation (Aliev and Azizov, 1977, Volodarskii et al., 1980, Sakharova et al., 1984, Khramova et al., 1988).

The investigation of peculiarities of photosynthetic carbon metabolism and its enzymes in cultivars of different productivity is of special interest as it determines the efficiency of the photosynthetic apparatus. In this respect, the results of an investigation of photosynthetic carbon metabolism, the activities of enzymes of primary carbon fixation, and photosynthetic rates in winter wheat genotypes differing in their productivity, photosynthetic properties, and sink source relationships are given in this work. Transport and distribution of assimilates have also been studied in wheat genotypes that differ in their photosynthetic properties and yield.

MATERIALS AND METHODS

The rich genefund, comprising several thousand wheat genotypes selected from both the ancient, aboriginal varieties of national selection and introduced from the world genefund, particularly, from *CIMMYT, ICARDA* and other International Centers, with contrasting photosynthetic traits, productivity and tolerance to water stress was created (Figure 1). All these genotypes were grown in field conditions in a wide area at the Absheron Experimental Station of the Research Institute of Crop Husbandry at the optimal regime of mineral nutrition and water supply under the guidelines established by the Department of

Plant Physiology and Biotechnology of the Research Institute of Crop Husbandry on the basis of long-term studies of morphophysiological traits of the varieties (Figure 2). The area of the inventory plot was 54 m^2; all experiments were replicated no less man three times.

Numerous winter wheat genotypes were the main targets of the research, and the most typical types are presented in this paper. The main parameters for selection of these genotypes were grain yield, plant phenotypic features (stem height, area and architectonics of the leaf surface, etc.), duration of the vegetation and other morphophysilogical traits, as well as drought resistance (Figure 3) (Aliyev and Kazibekova, 1977, Aliyev et al., 1982 ; Aliyev, 1983, 2002).

Due to leaves oriented under different inclination angles, plants create canopies with drooping (the angle is 30-40° from the vertical), semi-vertical (20-27°) and erect (10-18°) leaves. Varieties of intensive type are short-stemmed, leaves of vertical orientation, highly productive, and varieties of extensive type are long-stemmed with drooping leaves of horizontal orientation, low productive. Bread wheat (Triticum aestivum L.) varieties contrast in architectonics, Gyrmyzy gul and Azamatli-95, are short-stemmed (with the stem height of 85-90 and 60-75 cm, respectively), intensive type, with vertically oriented small leaves and high productivity (7-9 t ha^{-1}); Giymatli-2/17 is a short-stemmed (85-95 cm), intensive type, with broad drooping leaves and high productivity (7 t ha^{-1}), and Kansas-63323 is medium-stemmed (90 cm), with a small ear and small leaves, and grain yield of 3 t ha^{-1}.

Durum wheat (Triticum durum L.) varieties were used: Shiraslan-23 and Garagylchyg-2, are short-stemmed (78 and 82-85 cm, respectively), intensive type, with the vertical orientated leaves and potential grain yield of 6-8 t ha^{-1}; Sharg and Caucasus are medium-stemmed (110-120 cm), semi-intensive, with semi-vertical leaf arrangement and medium grain yield (4-5 t ha^{-1}); Oviachik-65 (CIMMYT) is short-stemmed (60-70 cm), with erect leaves and average grain yield of 6 t ha^{-1}; Gyrmyzy bugda and Sary bugda are long-stemmed (150-180 and 125-150 cm), extensive type, with drooping horizontal leaves and grain yield of 3 t ha^{-1}. Except Kansas-63323, Oviachik-65 (CIMMYT) and Caucasus (Krasnodar Agricultural Research Institute) all other varieties are of local selection (Figure 4, Table 1) (Catalogue, 2000; Aliyev, 2006).

The basic parameters of photosynthetic activity: leaf, stem and ear area, the rates of photosynthesis and photorespiration were determined during the ontogenesis.

In order to measure the rate of gas exchange in leaves of various layers and other assimilating organs, an infrared gas analyzer URAS-2T ("Hartman and Braun", Germany) with a short exposure of the whole plant in sowings in

$^{14}CO_2$ atmosphere (Figure 14) was used. The limits of measurements were 0.005-0.05% CO_2, error was ± 0.5% of the upper limit of the scale (Voznesensky, 1977; Aliev et al., 1996 b). CO_2 concentration in the analyzed air was recorded using an automatic recorder. The measurements were performed in an open-air flow system connected in the differential mode (Karpushkin, 1971).

The initial airflow was divided into two parts. One part passed through the air dehumidifier, filled with calcium chloride, through the filter, and then through the control cuvette of the gas analyzer. The other part passed through the leaf chamber, dehumidifier, filter, and then through the measuring cuvette. The airflow velocity through the entire system was adjusted using needle valves and rotometer. The gas analyzer recorded the difference in CO_2 concentration at the inlet and outlet of the leaf chamber. The rate of gas exchange in leaves placed in the leaf chamber was determined by the difference in CO_2 concentration and air velocity passing through the leaf chamber. For the measurements a hermetically sealed clip chamber with the area of 0.1 dm^2, which has two inlets and outlets for air flow, separately surrounding the upper and lower leaf surface, was used.

During the measurements the chamber was attached to leaves of different layers close to the stem maintaining their natural location and orientation, and exposed to sunlight until the gas exchange reached the steady-state level. The CO_2 concentration in air was determined close to the leaf chamber before each gas exchange measurement. Night respiration was determined using the above-mentioned equipment without use of a thermostat, in a steady night temperature.

Figure 1. Wheat genefund of the Research Institute of Crop Husbandry.

a

b

Figure 2. Experimental crop fields of the Department of Plant Physiology and Biotechnology of the Research Institute of Crop Husbandry.

a

b

c

Figure 3. (Continued).

d

Figure 3. (a) *Triticum aestivum* L. wheat varieties; (b) *Triticum durum* Desf. wheat varieties.

Figure 4. Wheat varieties developed at the Department of Plant Physiology and Biotechnology of the Research Institute of Crop Husbandry.

In the heat of the day a light filter SZS-24 (Voznesensky, 1977; Aliev et al., 1996 b) was used to prevent overheating of leaves in the chamber. Photorespiration was determined using two methods, in atmosphere without CO_2 and in atmosphere with reduced oxygen content (2%) (Šesták et al., 1971; Akhmedov, 1986). In the first case, after photosynthesis had reached the steady-state level the CO_2-lacking air was passed through the leaf chamber.

Table 1. Characteristics of the main modern wheat genotypes cultivated in Azerbaijan

No	Varieties	Origin	Parents	Years of cultivation	Cultivation areas	Main Features
1.	Mugan	Research Institute of Crop Husbandry	Created by an individual selection	1979	Uncovered rainfed of the republic	Winter durum wheat. Resistant to smut and rust, high-yielding, with high quality, protein content in grain 15.2-17.7%, glutens-28%.
2.	Shirvan-3	Research Institute of Crop Husbandry	Created by an individual selection from the sample *Leukurum*	1988	Uncovered rainfed in Nakhchivan area of the republic	Durum wheat. High-yielding variety (35-50 cwt/ha), resistant to sloughing, brown rust and smut, protein content 14.0-14.8%, glutens-27-30%.
3.	Garagylchyg-2	Research Institute of Crop Husbandry	Crossing local "Garagylchyg" cultivar with "Norin-10" by a multistage selection.	1988	In irrigated Shirvan-Garabag areas and rainfed areas of the republic	Durum wheat. High-yielding with best architecture, drought resistance is weak but it is resistant to rust diseases, powdery mildew and covered smut but may weakly be infected with loose smut
4.	Mirbashir-50	Research Institute of Crop Husbandry	Sort *Pobellon-67* from Mexico x local sort *Shark*	1988	Irrigated and rainfed areas of the republic	Durum wheat. High-yielding variety (55-60 cwt/ha) with high quality and vitrescence. Resistant to diseases, high pasta quality, protein content 12.5-16.0%, glutens 23.5-35.0%.
5.	Mirbashir-128	Research Institute of Crop Husbandry	Sort Bezostaya-1xC-271 from Pakistan	1995	In irrigated and rainfed areas of the republic	Bread wheat. High-yielding variety (60-70cwt/ha), protein content 10.8-14.7%, high baking quality, resistant to decumbency and diseases.
6.	Shiraslan-23	Research Institute of Crop Husbandry	Local sort Shark x Mexican sort Oviachik-65 with the subsequent selection by precocity	1995	In irrigated areas and in rainfed areas with green winter	Durum wheat. Low height, high-yielding (cwt/ha), precocious plant. Protein content 14.0-15.5%, glutens-28-30%, resistant to diseases and decumbency.

No	Varieties	Origin	Parents	Years of cultivation	Cultivation areas	Main Features
7.	Taraggi	Research Institute of Crop Husbandry	By hybridization of the Mexican sort K-290612 x Panonya from Yugoslavia	1996	In irrigated areas and in covered rainfed areas of the republic	Bread wheat. Middle height, resistant to decumbency and sloughing, high-yielding (70 cwt/ha), protein content 14.5-15.5%, glutens-30-35%.
8.	Turan	Research Institute of Crop Husbandry	Multiple selection from hybrid populations of durum wheat	1998	In irrigated areas	Durum wheat. High-yielding, high pasta quality, resistant to diseases.
9.	Azeri	Research Institute of Crop Husbandry	Sort Poloniya from Yugoslavia x Bezostaya	1998	In irrigated areas of Azerbaijan	Winter bread wheat. Middle height, high-yielding, high baking quality.
10.	Barakatli-95	Research Institute of Crop Husbandry	Gyirmyzy bugda x Garagylchyg-2	1998	Irrigated, foothill, rainfed areas	Durum wheat. High-yielding variety (55-60 cwt/ha), precocious, low-height, frost and drought tolerant, have high pasta quality of grain, high vitrescence, tolerant to diseases and decumbency.
11.	Giymatli-2/17	Research Institute of Crop Husbandry	Individual selection from genefunds of the International selection centers	1999	Irrigated and foothill areas	Bread wheat. High-yielding variety, with high quality, resistant to frost and diseases, not decumbent, high vitrescence.
12.	Alinja-84	Research Institute of Crop Husbandry	Shark x Vugar with the subsequent selection	1999	Irrigated areas and rainfed areas with green winter	Winter durum wheat. High-yielding variety, with high-quality, high vitrescence, high resistance to diseases and decumbency.
13.	Akinchi-84	Research Institute of Crop Husbandry	Individual selection from the Hungarian sample	1999	Irrigated areas and covered rainfed areas of the republic	Winter bread wheat. High-yielding variety (65-70 cwt/ha), resistant, not decumbent.

Table 1. (Continued)

No	Varieties	Origin	Parents	Years of cultivation	Cultivation areas	Main Features
14.	Azamatli-95	Research Institute of Crop Husbandry	Individual selection from the international nursery CIMMYT/CARDA	2005	Irrigated areas and areas of covered and uncovered rainfed of foothill regions of the republic	Winter bread wheat. Potential yielding is 80-90 cwt/ha. Tolerant to decumbency, diseases, drought. Characterized with high quality.
15.	Nurlu-99	Research Institute of Crop Husbandry	Individual selection from the international nursery CIMMYT/CARDA	2005	Irrigated areas and areas of covered rainfed of foothill regions of the republic	Winter bread wheat. Potential yielding is 80-85 cwt/ha. Tolerant to decumbency, diseases, sloughing. Characterized with high quality and precocity.
16.	Gobustan	Research Institute of Crop Husbandry	Individual selection from the international nursery CIMMYT/ICARDA	2007	Irrigated areas and areas of covered and uncovered rainfed of semi-drought and foothill regions of the republic	Winter bread wheat. Potential yielding is 70-80 cwt/ha. Tolerant to decumbency, diseases, drought. Characterized with high quality and precocity.
270.	Ruzi-84	Research Institute of Crop Husbandry	Individual selection from the local collection	2007	Areas of uncovered rainfed of semi-drought and foothill regions of the republic	Winter bread wheat. Potential yielding is 70-80 cwt/ha. Tolerant to decumbency, diseases, drought. Characterized with high quality and relative precocity.

The increase in CO_2 concentration at the chamber outlet is an indicator for the estimation of photorespiration. In the second case, after photosynthesis had reached the steady-state level the air with a reduced content of oxygen was blown into the chamber, and the obtained values of photosynthesis were measured. The value of photorespiration was determined as the difference between the photosynthetic values in low and normal oxygen content of the air. The gas analyzer, which was placed in a mobile laboratory, allowed multiple measurements in the sowings of different genotypes to be performed in a short time while keeping the high sensitivity of the facility in the field and maintaining the natural course of physiological processes in entire plants.

Photosynthetic carbon metabolism and utilization of main photosynthetic products were studied radiometrically under ambient CO_2 (0.03%) and O_2 (21%) (Aliev et al., 1996 b, c). In the experiments an open gas system was used for the $^{14}CO_2$ incorporation into various organs of plants. The required air reserve with radiolabelled carbon dioxide was prepared and kept under high pressure in 10-liter steel cylinder with a needle valve. The 50 cm^3 thermostated leaf chamber made of organic glass was used.

Transparent plastic bags were used for the $^{14}CO_2$ incorporation into the entire plants. The experiments were carried out under direct sunlight illumination. Initially, photosynthesis reached the steady-state level in an atmospheric airflow, then the radiolabelled air from the cylinder was blown through the leaf chamber with velocity of 1 l/min and CO_2 specific radioactivity of 1.000MBq/l. After 10 minutes of exposure, plants were removed from the leaf chamber and quickly fixed with boiling ethyl alcohol. Experiments were repeated 4-5 times. Radiochemical analysis of fixed plant material was carried out using the standard technique (Voznesensky et al., 1965; Keerberg et al., 1970). Water- and alcohol-soluble products were separated by two-dimensional paper chromatography. Measurement of the radioactivity of fractions and separate compounds was carried out using a scintillation counter SL-30 in dioxane-based standard scintillator. Radioactivity of fractions and separate compounds was calculated taking into account the self-absorption coefficient.

To investigate the transport and distribution of photosynthetic products under field conditions, $^{14}CO_2$ of ambient concentration and specific activity of 200 MBq/l was incorporated during photosynthesis into leaves of certain layers for 15-20 min, and then samples were prepared to determine their radioactivity. After a 20 min exposure, the chamber was removed from the plant; after 24 h or at the end of plant growth the plants were withdrawn from the soil, fixed with dry heat, dissected into individual organs and dried. The plant organs were weighed and then grounded for sample preparation (Aliev et al., 1996 a), and the radioactivity calculated per units of weight and per organ was measured. The radioactivity of the samples was measured using the end-window counter of SBT-13 in a layer of complete absorption. All experiments were replicated at least three times. Data in the tables represent meal values and their standard errors. Contribution of photosynthesis of the separate leaves, ear and other organs in protein synthesis in grain was defined by means of blocking photosynthesis of these organs and by exposition of all plants in sowings in $^{14}CO_2$ atmosphere (Voznesensky, 1977). In both varieties sink source relations were changed artificially. 10-20 samples of each variety were analyzed, all 7-layered leaves were removed, and leaves of the 8th layer and ear were not. In the other analysis 8-layered leaves were removed, leaves of 7th layer and ear were not. Next, a part of an ear was removed, but not the leaves. In the final analysis all 7- and 8-layered leaves were removed, and the full ear was left. Grain protein content was determined according (Cook and Evans, 1978). Extraction of different protein fractions was carried out by the standard method (Pleshkov, 1968). Determination of radioactivity of different protein fractions was conducted using a similar methodology of studying the metabolism of photosynthesis products. In order to determine the enzymes activities, the leaves were washed, after both ends were cut off they were homogenized by mechanical MPW-302 type disintegrator for 3 min in 0.05 M Tris-HCl buffer, pH 8.5, containing 1 mM dithiothreitol (DTT), 5 mM $MgCl_2$, 1 mM EDTA and 1% polyvinylpyrrolidone K-25 ("FERAK"). Homogenization of wheat ear elements was performed in a mortar after the ear had been divided into separate elements, i.e. awns, glume and grains. Homogenate was squeezed through the 4-layered gauze and centrifuged first for 10 min at 1000×g, then for 30 min at 5000×g at 4°C. The precipitate was removed, and the supernatant was decanted and immediately used for enzymes assays. The carbonic anhydrase activity was determined by an electrometric method according to the initial pH change during CO_2 dehydration (Rickli et al., 1964). The activity of RuBPC was determined spectrophotometrically ("ULTRASPEC", LKB, Sweden) at the 340 nm of optical density and 30ºC based on the quantitative determination of 3- phosphoglycerate (3-PGA) in the presence of phosphoglycerate kinase and glyceraldehyde phosphate dehydrogenase (Aliev et al., 1988, 1996 b, c). The reaction mixture contained 0.05 M Tris-HCl buffer, pH 7.8, 0.05 M $NaHCO_3$, 0.01 M $MgCl_2$, 0.005 M DTT, 0.01 M ATP, 0.25 mM NADH, 0.3 mM RuBP, 10 U of glyceraldehyde phosphate dehydrogenase, 10 U of phosphoglycerate kinase and 0.2-0.4 mg of the studied preparation. Control contained all components excepting NADH.

The activity of RuBPO was assayed by amperometric method (Romanova, 1980; Aliev et al., 1988, 1996 b, c). The reaction mix contained 50 mM Tris-HCl buffer, pH 8.6, 5 M $MgCl_2$, 0.5 mM RuBP, 1-3 mg of protein of preliminarily activated enzyme preparation. The enzyme was activated by incubation at room temperature for 5-10 min in the presence of 10 mM $NaHCO_3$ and 5 mM $MgCl_2$ at pH 8.6.

The activity of PEPC was determined by a spectrophotometric method (Romanova, 1980). Carbonic anhydrase activity was estimated electrometrically from the initial pH changes during CO_2 hydration (Wibur and Andersom, 1948). RuBP and PEP carboxylase activities were assayed spectrophotometrically, and RuBP oxygenase activity was measured after the amperometric method (Romanova, 1980). The highly purified carbonic anhydrase preparation was obtained from wheat leaves by our method (Aliev et al., 1989). Dialyzed protein fraction obtained by carbonic anhydrase purification, using precipitation at 76% saturation with $(NH_4)_2S0_4$ (Aliev et al., 1989) was used for RuBPC purification. The protein fraction was loaded into the column (2.5 x 100.0 cm) with *Sephadex G-200,* balanced with 0.05 M Tris-HCl buffer, pH 8.43, containing 0.001 M EDTA, 0.01 M dithiothreitol and 0.01 M NaCl. The same buffer was used for elution. Fractions possessing the RuBPC activity were combined and ultra-filtered in a helium atmosphere through *Ripor-4-25* membrane. The concentrated RuBPC solution was used for the experiments.

The rate of light-saturated electron transport was measured spectrophotometrically. Leaf assimilating area was measured using an automatic area meter “AAC-400” (“Kayashi” Delkon Co LTD, Japan). The specific leaf density was calculated as the ratio of its dry weight to the area.

Parameters of water regime were determined according to the procedure (Methodological guidelines, 1987, Boyer, 1995).

Protein concentration was determined according to the Lowry et al. (1951). The arithmetic mean and standard errors presented in the figures and tables were calculated based on the data from at least 4 biological replicates. The obtained data was statistically processed by standard analysis methods (Kaplan, 1970, Dospekhov, 1985).

RESULTS AND DISCUSSION

During 60 years of comprehensive investigations of photosynthesis, photorespiration and productivity of various wheat genotypes in natural conditions of cultivation, characteristics and parameters of photosynthetic activity of these genotypes in crop fields most closely correlating with plant productivity, have been established. The main emphasis was put on: 1) architectonics; 2) CO_2 assimilation; 3) leaf activity within the day and vegetation, and others.

According to our previous concept (Aliyev, 1974; Aliev and Kazibekova, 1979; Aliyev and Kazibekova, 1988), the optimum plant height and favorable leaf orientation in compact sowings contribute to the effective absorption of solar radiation and development of vegetative and economically valuable organs, i.e. activate those key links of productivity, which ultimately determine the high productivity of wheat variety of an "ideal" type (Aliyev, 1983).

Optimality of the plant growth is one of the main conditions for the better implementation of the photosynthetic functions. The first stage in solving the problem of grain high yield was detection of stem shortness, since long stem promoted to decumbency and decrease in the yield, especially with abundant irrigation, at high soil moisture, in use of a large amount of N fertilizers worsening also grain quality. Stem shortening favored to a good perception of the conditions stimulating plant development resulting in the formation of the high yield with higher ratio between grain and straw (40:60-50:50).

If considering stem shortness as a real way to increase yield, it should be noted that the stem height must not be less than fixed optimal values otherwise it results in the disturbance of the conditions of active functioning of the necessary leaf area, which is important in creation of the optimal seeding density during phases of earing, grain filling and normal photosynthetic function of 3-4 leaves. At the same time optimality of the growth is a prerequisite for another, no less important condition – alignment of plants in seeding with vertically oriented leaves.

Investigations of the yield and elements of the yield structure of the mixed sowing of varieties taken in different ratios – low-height, middle-height and high – showed that 63-83% of the total yield is accounted for high plants.

To use solar energy with feasible high efficiency, plants need a certain leaf area, which is 50,000-70,000 m^2/ha for wheat during the period of maximum development. However, actions organized with the aim of achieving high yield, frequently lead to the creation of an extreme area of the leaf surface, which exceeds 100,000 m^2/ha.

Experiments confirm that intensive wheat varieties, contrary to extensive ones, have the ability to form less leaf areas, which do not exceed maximal levels in conditions of the intensive agrotechnics. Comparative analysis of the leaf parameters of different genotypes detected a wide diapason of their changes (length 14-35cm, width-13-20 mm, area 15-55 cm^2, angle of leaf inclination from stem-10-70) and showed that in classic, extensive varieties leaves are mainly long and relatively narrow; while in modern intensive short-stemmed varieties leaves are mainly short and wide with large areas; genotypes with narrower, short and vertically oriented leaves also occur (figure 1).

The optimal values mentioned above, for assimilating surfaces of intensive genotypes, are reached namely at the expense of narrow, short, vertically oriented leaves with the rationally limited height.

The vertical orientation of leaves with a low variability of the growth parameters promote the efficient use of solar radiation energy, particularly with the high photosynthetic activity of different horizontal seeding layers, as well as to disease tolerance. Genetic conditionality of such leaf structures and their variability allow greater selection of purposeful works.

The next main characteristic is a specific surface density (SSD) value. Investigations showed that the value inherent in genotypes depending on factors determining photosynthetic activity is not always constant (Aliyev and Kazibekova, 1988). The SSD variation in upper leaves of some durum wheat genotypes is 0.35-0.53 g/m^2 during the earing phase – which is one of the ontogenesis periods responsible for yield – and besides, extensive varieties with high leaf indexes are characterized with significantly lesser values of this parameter than intensive varieties. Samples with rather high SSD occur in genotypes having small leaves. In total, with the leaf disc orientation depending on the leaf group and the illumination degree, these signs are reflected in the ultra structure of chloroplasts and in the middle-sized photosynthetic units, which in their turn are crucial for the effective conversion of the absorbed energy.

Investigations of photosynthesis as a donor of assimilates in interaction with the growth functions during the plant development process showed the importance of the duration of the donor-acceptor way in the assimilate transportation. The role of the distance from the wheat last leaf to the ear is indubitable. It was believed that the short distance provided a fast assimilate shift from leaves to ear and small expenditures of organic compounds for respiration that promoted the intensified accumulation of the compounds in grain. Nevertheless the detailed study of the problem in different genotypes showed the absence of the positive correlation between the small distance from the last leaf to ear and yield. The reason is that the given part of the stem being long (definitely within certain limits) prevents the possibility of severe diseases and promotes a better supply of plants with light and CO_2, etc.

because of the intensity of photosynthesis and an active participation in forming yield. The distance from the last leaf to the ear in high forms of wheat averages 30-50 cm and in low-height forms-12-20 cm or 30-35 and 20-25% of the total length of the stem, respectively. Low-height forms provide more active sink strength of the ear.

Duration of functioning of all leaf groups during vegetation is also important for providing information about the morphogenetically most important photosynthetic activity period for every leaf with certain adjustments. At the same time a correlation between chloroplast photochemical activity and productivity during the crucial period of earing-grain filling has been identified (Aliyev and Azizov, 1977). So the high photochemical activity of chloroplasts could be assumed for one of the signs of ideal wheat.

The complex of the parameters of the final yield also involves amounts of the productive stems per seeding unit, average amounts of grains in ear and grain mass; simple increases there do not always lead to a high yield. However, having forms with big, well-filled ears could facilitate improvement of the genetic potential of varieties and high yield.

Evaluation of the breeding material and direction of the studies, according to these generalized principles of the use of photosynthetic signs, allowed shortening terms of the breeding works and create varieties (Figure 4) close to the presented ideal wheat type. Its parameters: height-60-80 cm; tops of stems are aligned; ears are straight, leaves are narrow, short with high SSD and vertically oriented. The optimal distance from the last leaf to the ear for durum wheat is 20-25% of the total stem length. Leaves, stem and ears possess the maximum photosynthetic activity, providing high productivity. This variety is resistant to decumbency and many diseases, well tillered, easily ingests high dosages of N and with good farming practices, at 4-5 m/ha seeding rate the grain yield is 70-100 cwt/ha.

Analysis of various wheat genotypes with different values of photosynthetic traits and productivity in conjunction with a range of environmental factors, including mineral nutrition, water, light, etc. showed the wide range of CO_2 assimilation variability in ontogenesis, depending on the morphophysiological peculiarities of genotypes and their sink source relations (Aliev and Kazibekova, 1995).

The rate of its integral photosynthetic process and also its enzymatic and metabolic activity changes during the leaf life cycle. Obtained results showed that the studied genotypes also differed significantly in the level of photosynthetic carbon metabolism (Jahangirov, 1987, Aliev et al., 1996 b).

The photosynthetic rate has increased steadily during flag leaf development and reached its maximum value in the earing stage (Figure 5). Two peaks of CO_2 exchange, in the most active leaves, were observed during ontogenesis: the first peak at the beginning of stalk emergence and the second one at the beginning of flowering. It is important to search genotypes with no or less evident decrease of the photosynthetic rate in this period. High productive genotypes possess a higher rate of CO_2 assimilation throughout the whole lifespan of their flag leaves up to the end of the grain filling stage. This pattern of carbon dioxide exchange at the end of the growth period in different genotypes may be due to early death of the lower leaves of extensive varieties. However, in the intensive genotypes, in comparison with extensive ones, the decrease in CO_2 assimilation in the flag leaf was counterbalanced by the ac active CO_2 assimilation in the lower leaves. High productive varieties significantly exceeded (up to 1.5 times) low productive ones by means of the photosynthetic rate of upper leaves (flag leaf and the second leaf from the top).

In intensive varieties with the best architectonics, the rate of photosynthesis is higher in the morning, and especially in the evening hours, and afternoon depression of photosynthesis occurs later and is much weaker than that of semi-intensive and extensive varieties. The total photosynthesis during the daytime is higher in high productive varieties than in low productive ones.

The study of the dynamics of photosynthetic rate in ontogenesis showed the availability of multiple peaks specific for genotypes depending on the crop architectonics at the period, when the rate of photosynthesis was determined.

It was found that the rate of photorespiration is higher in intensive varieties than in extensive types. The rate of photorespiration reaches its maximum in the earing stage, and then decreases (Figure 6). In high productive varieties photorespiration remains at a high level during the periods from stalk emergence up to grain formation and filling than in the low productive varieties. Along with an increase of the photosynthetic rate, the rate of photorespiration increases too. Photorespiration follows the course of photosynthesis; the more intense is photosynthesis, and the more intense is photorespiration. The daily course of photorespiration, in general, coincides with the daily course of leaf photosynthesis at various stages of plant development. However, photorespiration begins at later hours than photosynthesis, and ends earlier; the curve of its course is unimodal.

The rate of CO_2 assimilation in leaves and other photosynthesizing organs is determined by the size and structure of these organs, plant architectonics both at the individual cultivation and in crop fields, by the sink source relations genetically stipulated in each genotype, taking into account all affecting factors (Aliyev, 2001 a).

Wheat genotypes are characterized by following parameters: leaf length, width, surface area, specific leaf density, leaf inclination angle and orientation, that create an optimal architectonics more favorable for turbulence and high rates of CO_2 assimilation, and possibly for long-term photosynthetic activity of all leaves, ears and other non-leaf organs during grain formation.

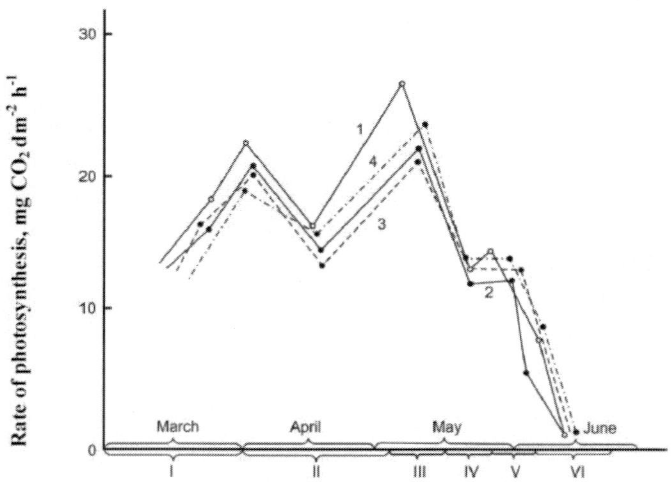

Figure 5. Ontogenetic changes in the rate of CO_2 assimilation in different wheat genotypes: 1 – Oviachik-65, 2 – Sharg, 3 – Gyrmyzy bugda, 4 – Caucasus. I – stalk emergence, II – earing, III – flowering, IV – grain formation and filling.

The genotypes with two or three times less leaf areas than that with broad leaves produce a similar or even greater grain yield. Genotype Gyrmyzy bugda with flag leaf area of 28 cm^2 yields up to 3 t ha^{-1}, and those of 18-19 cm^2 – 7-9 t ha^{-1}. In highly productive genotypes with grain yield ~ 7-9 t ha^{-1}, the flag leaf areas differ almost three times. The studied varieties of winter wheat exhibited significant differences in the rate of flag leaf CO_2 assimilation. The highest values were detected in high productive genotypes Gyrmyzy gul and Azamatli-95, whereas the lowest CO_2 assimilation was characteristic for genotype Gyrmyzy bugda. Vertically oriented small leaves create optimal architectonics, and probably promote a relatively high CO_2 assimilation of photosynthesizing leaves of all layers during grain filling. A comparative study of the photosynthetic rate of wheat genotypes with contrast architectonics during the day showed that the diurnal changes in the photosynthetic rate of leaves of all layers and genotypes are characterized by a double-peaked pattern indicating a drastic increase in the photosynthetic rate in the morning and decrease in the evening (Figure 7). Photosynthesis in the flag leaves began early in the morning (at 6 a.m.), increased during sunrise, reaching its maximum level at 11 a.m. Then photosynthetic rate declined to its lowest at midday. After midday depression of photosynthesis, the second peak was observed at 5 p.m., which correlates with increased photosynthesis. Not all genotypes with small leaves are highly productive and not all genotypes with broad leaves are high or low productive. Genotypes with broad leaves and high yield require sufficient water supply. In order to better understand the correlation between rates of CO_2 assimilation, photorespiration and productivity, the consideration of basic parameters of plant architectonics is also essential (Figure 7). Obtained data indicates that genotypes with vertically oriented short and narrow leaves (20-30 cm^2), high specific leaf density (SLD) – 600 mg/100 cm^2, with stable and long-term intensive CO_2 assimilation (30-40 mg dm^{-2} h^{-1}) and a high tolerance to water stress yield up to 10 t ha^{-1}. The high rate of CO_2 assimilation is not accompanied by a low rate of photorespiration. For high productive genotypes, the high values of photorespiration are common. Genotypes with grain yield of 7-9 t ha^{-1} possess high rates both of CO_2 assimilation and photorespiration at corresponding architectonics. Genotypes with moderate (4-5 t ha^{-1}) and low (3 t ha^{-1}) grain yield have relatively low rates of CO_2 assimilation and photorespiration.

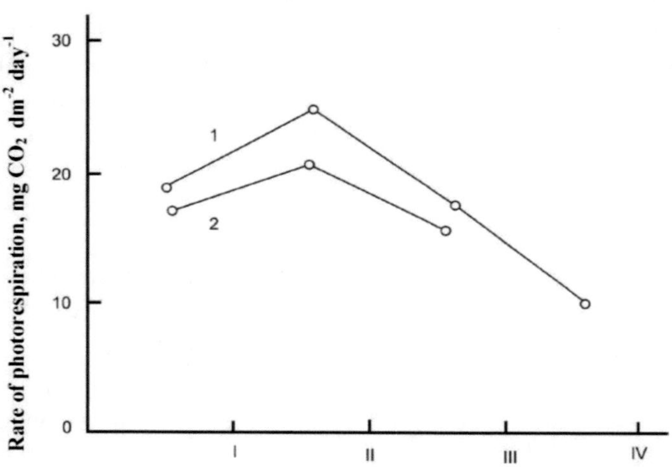

Figure 6. Ontogenetic course of total photorespiration in high productive (Oviachik-65 – (1)) and low productive (Gyrmyzy bugda – (2)) genotypes: I – stalk emergence, II – earing, III – flowering, IV – grain formation and filling.

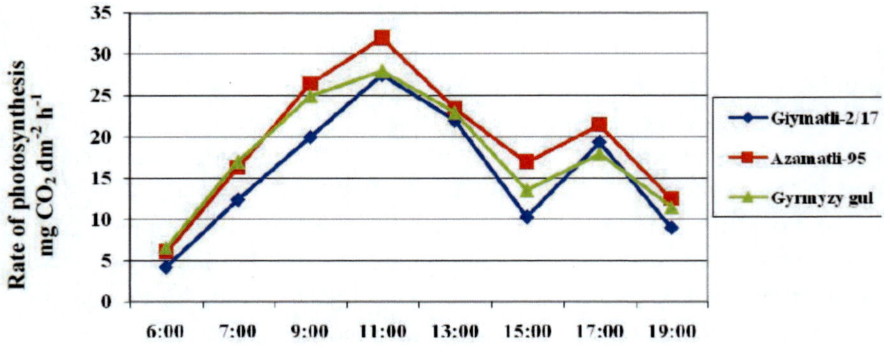

Figure 7. Diurnal course of CO_2 gas exchange of flag leaves of wheat genotypes with different architectonics at the milk ripeness stage.

Gas exchange data closely agrees with the measured enzymes activities directly involved in CO_2 fixation. The activity of RuBPC and carbonic anhydrase in highly productive, intensive wheat cultivars was increasing monotonously from the beginning of flag leaf formation to reaching its maximum at the end of anthesis and then it decreased toward the end of vegetation. By contrast, the activities of these two enzymes of high-growth extensive wheat decreased quickly after reaching its maximum (Figure 8).

The activities of RuBPO and RuBPC were higher in highly productive wheat genotypes than in the low productive ones. The variation of RuBPO activity in the course of the flag leaf development was similar to that of RuBPC.

Moreover, the ratio of the activities of RuBPC/RuBPO during the ontogeny of flag leaf of the genotypes studied was maintained at a similar level (Garagylchyg-2 - 19.01+1.36, Shiraslan - 23 - 18.66±0.898, Gyrmyzy bugda - 16.02±1.99, Sary bugda - 16.2+1.07) (Figure 9). Nevertheless, the variation of the RuBPC/RuBPO activity ratio during the flag leaf formation was associated with a certain genotype difference.

The PEPC activity remained sufficiently high at all the stages of the flag leaf development of the genotypes studied. The enzyme activity increased during the flag leaf formation and then it decreased. The PEPC activity was higher in highly productive wheat genotypes than in the low productivity ones.

A sharp variation in the RuBPC activity of different elements of the ear depends on the extent of its development and on the plant genotype. In comparison with other ear elements, the glume of both intensive and extensive genotypes is characterized by a higher carboxylase activity.

At the beginning of grain formation the RuBPC activity in the ear glume of the intensive genotype was also higher than in the extensive one.

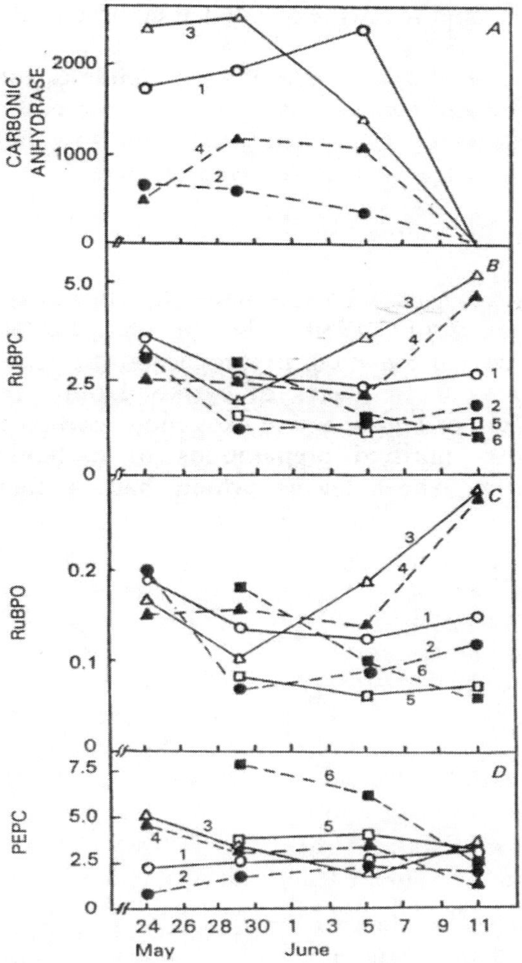

Figure 8. Variations in activities [mmol m-^{2}(leaf) s-^{1}] of carbonic anhydrase *(*A*)*, RuBPC (B), RuBPO (C) and PEPC (D) in ontogeny of flag leaf of wheat genotypes (1*)* Garagylchyg-2, (2) Shiraslan-23, (3) Gyrmyzy bygda, and (4) Sary Bugda. *The arrows* indicate the beginning of ear emergence. With highly productive genotypes the beginning of ear formation falls on May 6, the beginning of anthesis - on May 14 through 15; the beginning of grain maturation - on May 20 through 22; the beginning of milk ripeness - on June 4 through 6; the beginning of wax ripeness - on June 20 through 21. With low productive genotypes the beginning of ear formation falls on May 12; the beginning of anthesis - on May 18 through 20; the beginning of grain maturation - on May 25; the beginning of milk ripeness on June 12; the beginning of wax ripeness - on June 24.

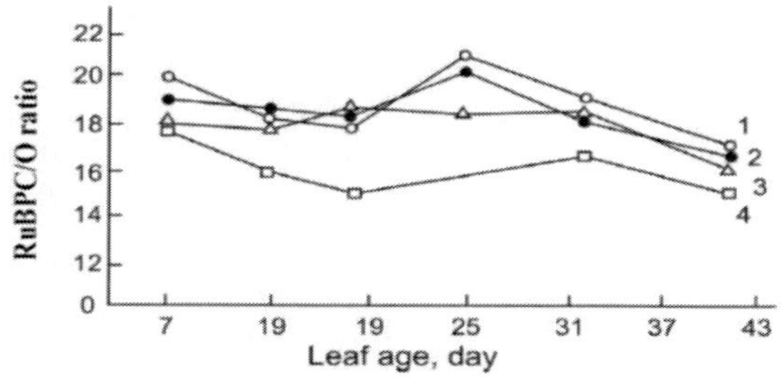

Figure 9. Changes in the ratio of RuBP carboxylase to oxygenase activities in flag leaf ontogenesis of different wheat genotypes: 1 – Garagylchyg-2, 2 – Shiraslan-23, 3 – Gyrmyzy bugda, 4 – Sary bugda.

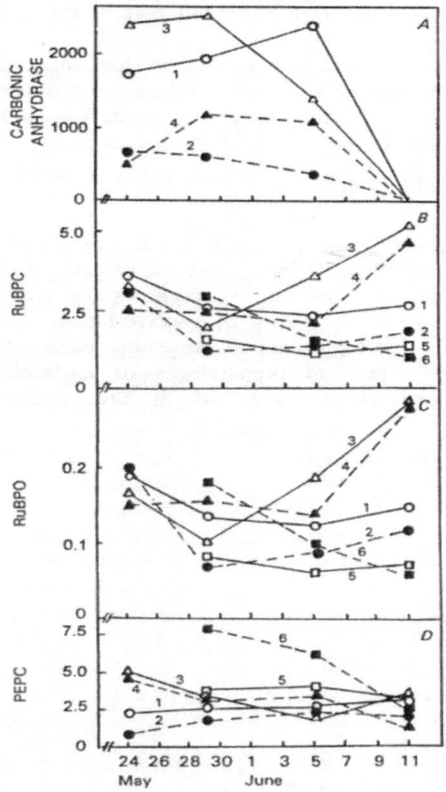

Figure 10. Variations in activities [mmol kg^{-2}(protein) s^{-1}] of carbonic anhydrase *(A),* RuBPC *(B),* RuBPO (c) and PEPC *(D)* in ear elements of wheat genotypes during grain maturation, *i.e.* (7) awn, *(3)* glume, and (5) grain of low-stature highly productive Shiraslan-23 genotype; (2) awn, *(4)* glume, and (6) grain of high growth low productive Sary bugda genotype.

However, later (up to a definite moment) this activity, calculated per unit of protein, decreased in the intensive genotypes but remained the same in the extensive ones. Later, during grain maturation, a significant increase in the glume RuBPC activity took place both in the intensive and extensive genotypes (Figure 10). During all periods of ear formation and grain maturation the ear awns of the intensive genotype had a higher activity of RuBPC than those of the extensive genotype.

The RuBPC activity in grain of the low productive genotype was higher than that of the highly productive one only at the beginning of grain formation. A gradual decrease in the enzyme activity during grain formation was observed in both genotypes. During grain formation the activities of RuBPC and of RuBPO in different ear elements changed in parallel, similar to the flag leaf.

From the end of anthesis and during the grain maturation a gradual decrease in the carbonic anhydrase activity took place in the awn of the low productive genotype. The enzyme activity increased significantly in the awn of the highly productive genotype, reached its maximum in the grain maturation phase, and then decreased sharply to the level observed in the low productive genotype. In both genotypes an increase in the glume carbonic anhydrase activity took place during the first days of the reproductive phase, then a sharp decrease in this activity occurred in the highly productive genotype. The glume carbonic anhydrase activity of the low productive genotype in comparison with the high productive one remained at a similar level for some time and then decreased (Figure 10). The carbonic anhydrase activity in all ear elements of the highly productive genotype significantly exceeded that of the ear elements of the low productive one.

Unlike PEPC, the activity of which decreased gradually in all the ear elements, the activities of RuBPC and RuBPO in the glume and awn of both genotypes increased during grain maturation. The activity of these enzymes in the glume of the low and highly productive genotypes, in comparison with the awn, increased considerably.

The interaction of carbonic anhydrase and RuBPC in cells was found by Bird et al. (1980), Pronina and Semenenko (1984), and Finis (1986). One of the possible functions of plant carbonic anhydrase is its participation in preparing the non-hydrated CO_2 molecule, *i.e.* a RuBPC substrate in places of carboxylation. To confirm such a possibility we studied the effect of highly active exogenous carbonic anhydrase on RuBPC *in vitro.* The highly purified preparations of carbonic anhydrase and RuBPC were obtained from wheat leaves, which had neither carboxylase nor carbonic anhydrase activity.

Our results showed that exogenous carbonic anhydrase stimulated the RuBPC activity. The RuBPC activity in the reaction medium increased with the increase in the carbonic anhydrase activity. However, as soon as a certain ratio of the activities of these enzymes was attained, the RuBPC activity shifted to the plateau, and a further increase in the carbonic anhydrase activity in the reaction medium did not bring about any increase in the carboxylase activity at the pre-set HCO_3^- concentration. The dependence of RuBPC activity on the HCO_3^- ion concentration in the absence of carbonic anhydrase had an S-shape form that vanished in the presence of carbonic anhydrase.

During the flag leaf development the activity of carbonic anhydrase and that of RuBPC in the highly productive wheat genotypes studied changed in parallel. Majeau and Coleman (1994) studied the relationship between the expression of carbonic anhydrase and RuBPC during pea development in various tissues and cultivars. Although the absolute levels of activity and the quantity of transcripts of these two enzymes may vary considerably during plant development, a relatively constant relation between the activities and transcript quantity of carbonic anhydrase and RuBPC, which can be observed during greening, is maintained in mature and senescent photosynthetic tissues. In different cultivars of pea, where the quantities of RuBPC and carbonic anhydrase vary, the ratio of carbonic anhydrase and RuBPC remains constant as well (Majeau and Coleman, 1994). Also our results proved the concerted action of carbonic anhydrase and RuBPC of wheat leaves under different environmental conditions.

The increase in the cooperative coefficient in the presence of carbonic anhydrase may be attributed to the increase in concentration of non-hydrated $C0_2$, *i.e.* the substrate and allosteric factor of RuBPC (Miziorko, 1979). Carbonic anhydrase accelerates the reactions of $HC0_3$--dehydration, increases the CO_2 concentration in the places of carboxylation, and thus contributes to the effective work of RuBPC in the cell. The high activities of carbonic anhydrase and RuBPC in highly productive wheat genotypes play a significant role in maintaining a high CO_2 assimilation.

So, there is no close correlation between the activities of RuBPC and carbonic anhydrase in all the ear elements in comparison with the flag leaf of the same wheat genotypes. This can be probably attributed to the peculiarity of the photosynthetic process in ear elements.

As stated before, the highly productive genotypes studied have a more intensive $C0_2$ absorption during the whole flag leaf ontogeny, except for the terminal stage of grain maturation (Guliev et al., 1985). A close correlation between the CO_2 assimilation and the activity of basic carboxylases was found. The highly productive genotypes had higher values of these indices than the low productive ones. On the other hand, the genotypes studied manifested significant differences not only at the level of primary photochemical reactions but also at the level of photosynthetic carbon metabolism, transport and distribution of photo-assimilates in various parts of wheat (Kerimov et al., 1987).

Our highly productive genotypes were characterized by high RuBPO activity and high rate of CO_2 evolution during photorespiration. Thus the studied modern low-growth genotypes with a favorable architectonics to create an optimal sowing structure enabling effective solar energy utilization had both a high rate of photorespiration and a higher photosynthetic rate. The products of glycollate metabolism can either be used in sucrose synthesis or transferred. Thus under certain conditions they can provide an active transport of photosynthates and at the same time create appropriate conditions to maintain photosynthesis at a high level.

Similar results proving the constant RuBPC/RuBPO ratio were obtained for spring and other cultivars of winter wheat (Thomas et al., 1978), for sea and freshwater algae (Kalinkina et al., 1981), for sunflower genotypes (Lloyd and Canvin, 1977), and for some cotton cultivars (Rasulov et al., 1983). The ratio remains constant throughout the plant growth.

The time of a sharp increase in the activity of both enzymes in the glume of both wheat genotypes coincided with the decrease of these activities in the flag leaf. Thus during the period of grain maturation the ear elements, and the glume in particular, participate actively in the process of CO_2 assimilation.

Thus, the results obtained using various methods on different plant genotypes showed that RuBP carboxylase and oxygenase activities change in parallel in the course of the plant breeding process. Each wheat genotype was characterized by a definite value of RuBPC/O ratio and its variation under different influences (in particular, under artificial violation of the sink source relations) was temporary. A similar conclusion can be drawn on the basis of measurements of the components of carbon dioxide exchange in different wheat varieties; accordingly, the rates of net and true photosynthesis were highest in Garagylchyg-2. The rate of CO_2 release in the light was 11.4 and 8.7 mg CO_2 (dm^2 h) in Garagylchyg-2 and Gyrmyzy bugda, respectively (Figure 11). As mentioned above, the rate of CO_2 release in the light due to dark respiration was similar in all varieties. However, the rate of CO_2 release due to photorespiration was higher in Garagylchyg-2 (8.7 and 6.2 mg CO_2 dm^2/h) in Garagylchyg-2 and Gyrmyzy bygda, respectively. The ratio of the rate of CO_2 release in the light due to photorespiration to the net CO_2 assimilation rate was similar in both varieties and averaged about 22.0 and 21.2%, respectively, which indicated a positive correlation between the rates of photosynthesis and photorespiration.

Features of photosynthetic carbon metabolism, as part of metabolism in plants, are significant for productivity, which is compounded by the passing of complex processes of gradual transformation of photosynthesis products - from CO_2 assimilation to the generation of end products. In this regard, the study of metabolism of photosynthesis products under natural and modified donor-acceptor interactions during leaf ontogeny is of particular interest, since the donor ability of leaves and other organs varies with age.

In order to facilitate the performance of this task we have studied in detail the kinetics of $^{14}CO_2$ incorporation into the products of steady-state photosynthesis during the flag leaf ontogeny only in Garagylchyg-2 variety, as an outstanding representative of high-yielding intense variety.

Diagrams of $^{14}CO_2$ incorporation in natural sink-source relationships in different photosynthesis products in Garagylchyg-2 variety during the flag leaf ontogeny are shown in Figure 12-13.

It appears from the figures that assimilation of $^{14}CO_2$ is linear regardless of the flag leaf age. This indicates the fixed rate of photosynthesis during the exposure of the examined leaf in the medium with labeled carbon dioxide, which also indirectly indicates that ambients have slightly changed throughout the entire experiment.

Incorporation of $^{14}CO_2$ in steady-state photosynthesis products demonstrates that pool of sucrose, malate and aspartate have no saturation even at 20-min exposure in the $^{14}CO_2$ medium regardless of the flag leaf age, while starch, PGA + PEs, alanine, glycine + serine and glycerate are characterized by a rich pool of metabolites, which accurately correlate with the data obtained from other C_3 plants (Mokronosov, 1978).

The kinetic parameters of carbon metabolism during ontogeny of flag leaf obtained on the basis of $^{14}CO_2$ incorporation dynamics in different products are shown in figure 14-15. Evidently, assimilation of CO_2 in the flag leaf ontogeny does not occur at equal rates. Thus, the intensity of CO_2 assimilation in flag leaf of Garagylchyg-2 increases and reaches its maximum (12.5 μmol $CO_2/dm^2 \cdot min$) at 20 days of age, which corresponds to the earing stage, then decreases in the flowering stage and remains practically constant up to the milk ripeness stage (age of 45 days).

Dynamics of growth of the flag leaf area showed that it reaches the maximum CO_2 assimilation in the same phase as its maximum area, i.e. in the earing stage. The same results were obtained earlier in the study of the rate of photosynthesis and photochemical activity of chloroplasts in the ontogeny of other wheat varieties (Aliyev and Azizov, 1977; Aliyev and Kazibekova, 1979 a, b, Aliyev et al., 1996a). As it can be seen from figure 15, the rate of sucrose synthesis changes in the same way as the CO_2 assimilation rate during the ontogeny. So with the intensity of CO_2 assimilation, the rate of sucrose biosynthesis increases during processing from 10 to 20 days of age and amounts 7.1 to 11, 3 μmol $CO_2/dm^{2}\cdot min$. In the flowering stage (age of 30 days), it reduced to 6.2 μmol $CO_2/dm^{2}\cdot min$ and remained nearly constant up to 45 days of age (milk ripeness). Comparison of the CO_2 assimilation rate and the sucrose synthesis rate shows that they change proportionally to each other during the ontogeny.

The rate of malate, aspartate and alanine synthesis in 10-day-old leaves is 1.04, 0.78 and 0.67 μmol $CO_2/dm^{2}\cdot min$, respectively (Figure 14). During the milk ripeness stage (age of 45 days) it decreases up to 0.35, 0.27 and 0.14 μmol $CO_2/dm^{2}\cdot min$, respectively. In relative units the synthesis of these compounds in 10 day old flag leaves is 12.8, 8.7 and 8.2% of the CO_2 assimilation rate, respectively, and 5.0, 3.6 and 2.8%, respectively, at 45 days of age.

Biosynthesis of starch differs from other compounds by greater stability and amounts to about 0.9 μmol $CO_2/dm^{2}\cdot min$ in leaf ontogeny, i.e. 8.5% (Figure 15), and the sucrose/starch ratio varied from 9 to 14. This pattern of change in the direction of biosynthesis of substances is primarily due to the fact that from 10-15 days of age the flag leaf area is very close to the maximum, and it is an active assimilate donor. A simultaneous increase in the rate of synthesis of carbohydrate compounds and a decrease in the synthesis of non-carbohydrate compounds was obtained for other C_3 plants as well (Mokronosov, 1972).

It should be noted that one of the reasons that impact the direction and pattern of metabolism is a natural change in the activity of carboxylation enzymes in the ontogeny of leaves. After reaching the maximum leaf area the RuBP carboxylase activity reduces more significantly than the same of PEP carboxylase (Guliyev et al., 1985). As it appears from Figure 15, the rate of biosynthesis and the total value of the glycine + serine pool increase in the earing stage, then decrease by the flowering stage, and remains nearly constant up to the milk ripeness stage, which is also identical to the change in the total CO_2 assimilation in leaf ontogeny (Figure 14).

Based on the fact that glycine and serine are produced by the carbon photorespiration pathway during photosynthesis in natural concentration of CO_2 and O_2 (Keerberg et al,, 1970), in our experiments, performed also under natural concentrations of carbon dioxide and oxygen, glycine and serine, specific metabolites of the glycolate pathway are produced as a result of photorespiration carbon metabolism. Therefore, the close relationship between the rate of biosynthesis of glycine+serine and the CO_2 assimilation rate indicates that processes of photosynthesis and photorespiration, and carboxylase and oxygenase activity of RuBisCo as well, are proportional to each other in wheat leaf ontogeny, which accurately correlates with the data obtained for winter wheat as well as for cotton and soya (Rasulov and Asrorov, 1985; Aliyev et al, 1992; Aliyev et al, 1996a).

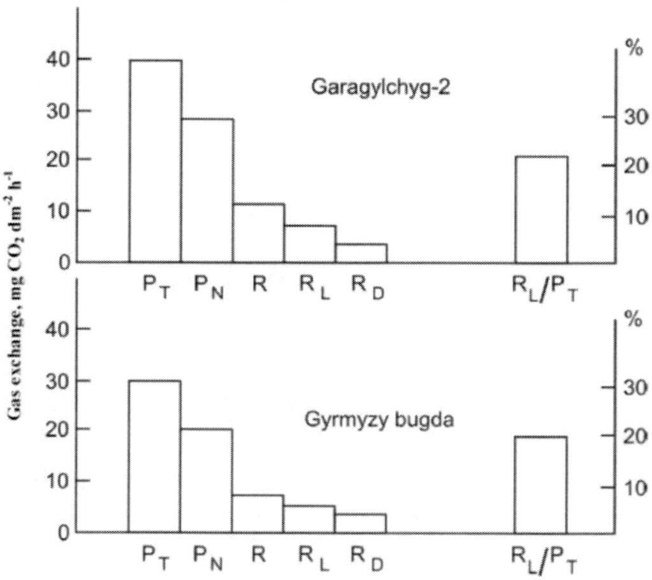

Figure 11. Components of CO_2 exchange in the wheat varieties Garagylchyg-2 and Gyrmyzy bugda (mg CO_2 dm^{-2} h^{-1}): P_T – true photosynthesis; P_N – net photosynthesis; R – CO_2 release in the light; R_D – CO_2 release in the light due to dark respiration; R_L – CO_2 release in the light due to photorespiration; R_L/R_T – rate of photorespiratory CO_2 release to true photosynthesis (%).

Good correlation of the data obtained by different methods and in different crops testifies that the photosynthesis/photorespiration ratio in leaf ontogeny remains sufficiently constant. With growing flag leaf of wheat the biosynthesis of non-carbohydrate compounds reduces and the carbohydrate metabolism direction (sucrose/starch ratio) rises. Biosynthesis of the glycolate pathway products (glycine + serine) is more dependent on the intensity of CO_2 assimilation in leaf ontogeny herewith. Diagrams of ^{14}C incorporation of the products of photosynthetic metabolism during earing and milk ripeness stages in other varieties of wheat (Gyrmyzy bugda and Sharg) are shown in Figures 16-18.

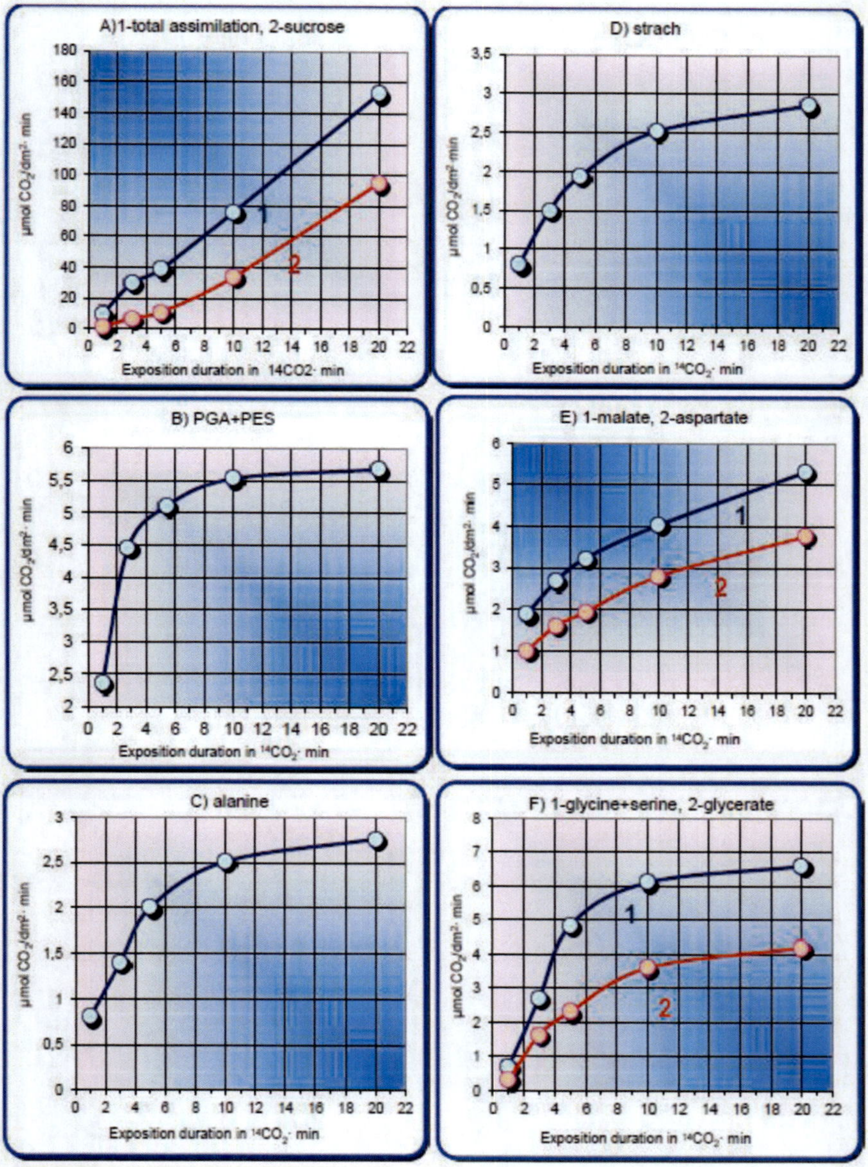

Figure 12. $^{14}CO_2$ incorporation in the products of photosynthesis in 10 days old flag leaf of Garagylchyg-2 (µmol $CO_2/dm^2 \cdot min$).

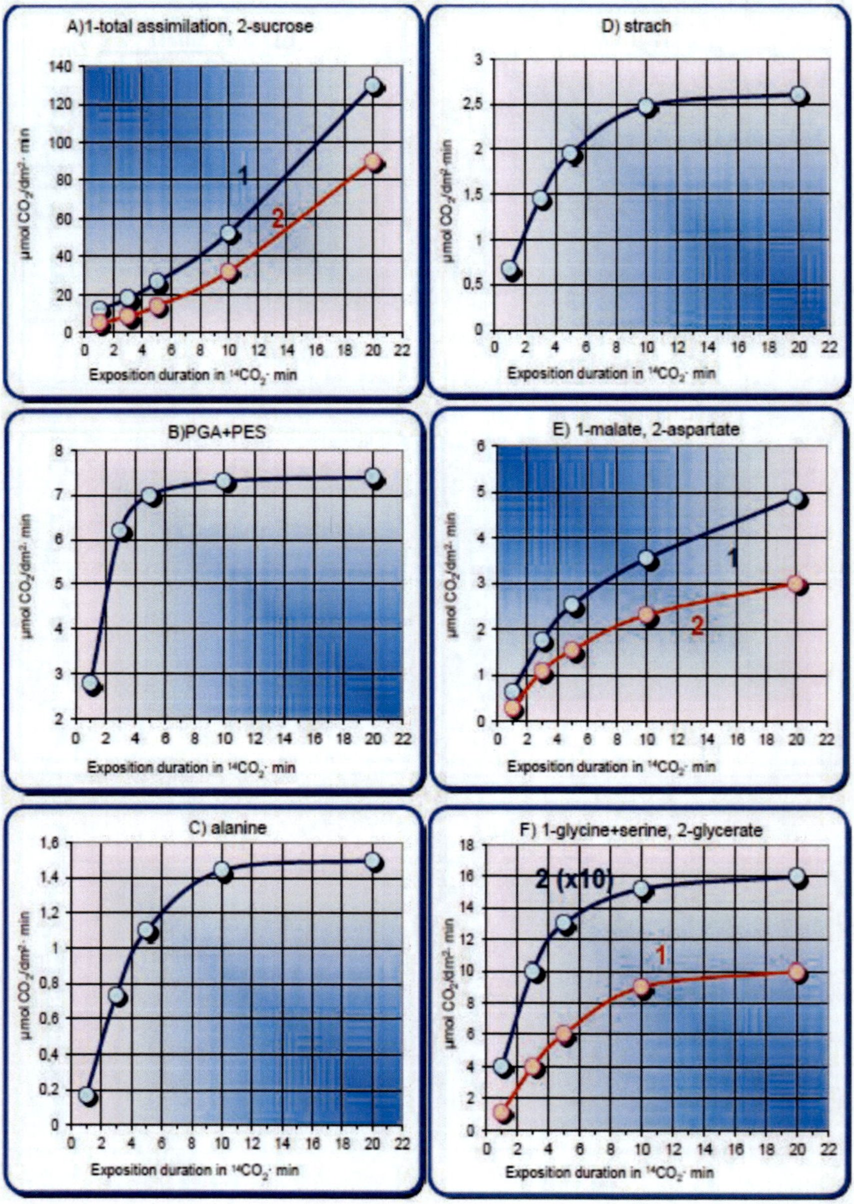

Figure 13. ^{14}C incorporation in the products of photosynthesis in 30 days old flag leaf of Garagylchyg-2 (μmol $CO_2/dm^{2\cdot}min$).

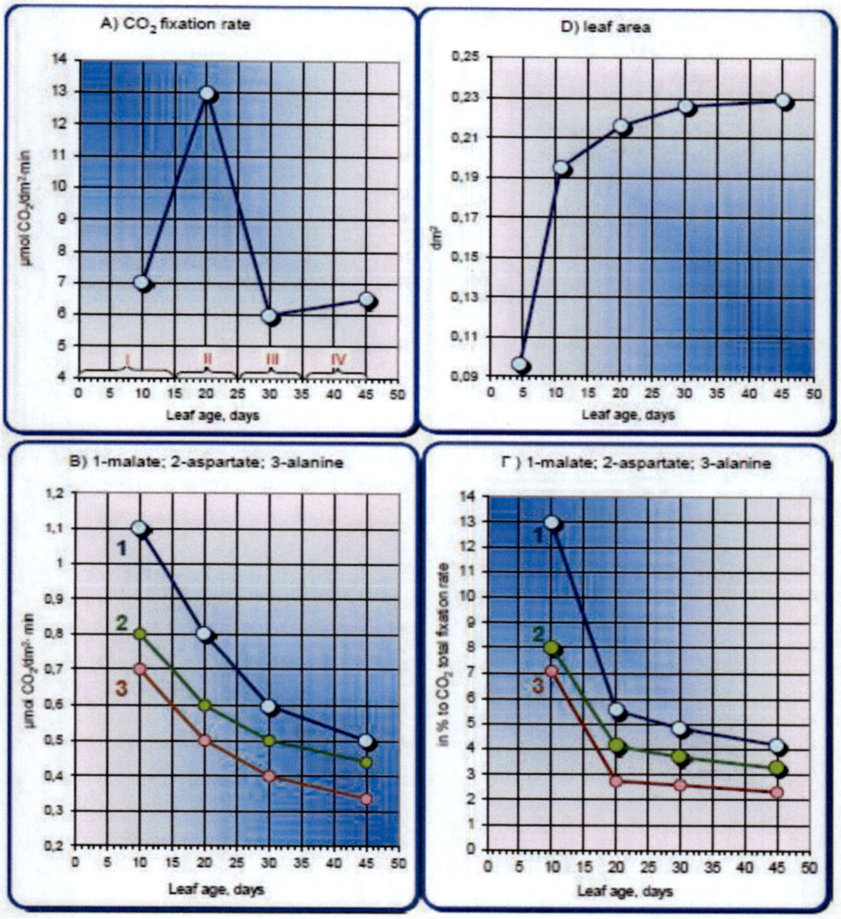

Figure 14. Biosynthesis of malate, aspartate and alanine in flag leaf ontogenesis of wheat variety Garagylchyg-2: I – stalk emergence, II – earing, III – flowering, IV – grain filling. A – rate of CO_2 fixation; B – leaf area; C – malate (1), aspartate (2), alanine (3); D – the same as C (% of the total CO_2 fixation rate).

In these varieties PGA + PES, alanine, glycine + serine, glycerate and starch have permanent pool saturation as a result of a 20-minute exposure of flag leaf as opposed to malate, aspartate and sucrose similarly to Garagylchyg-2.

Table 2 reflects the kinetic features of the photosynthetic carbon metabolism, calculated by diagrams of inclusion of ^{14}C in the photosynthesis products.

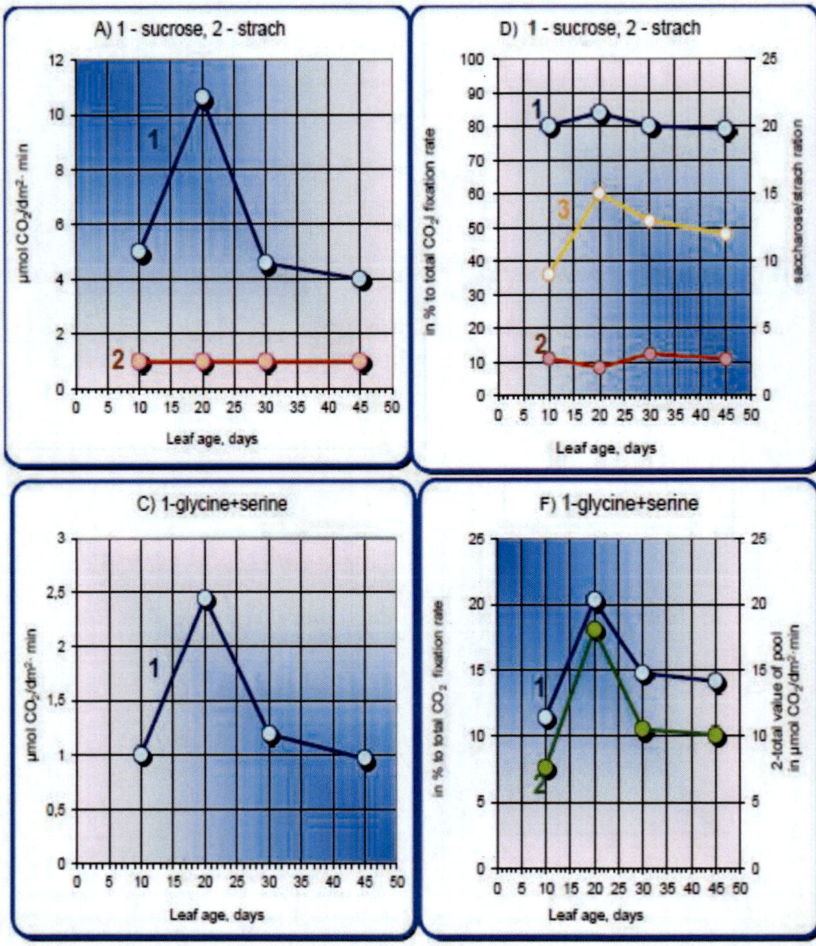

Figure 15. Biosynthesis of sucrose, starch and total glycine-serine in flag leaf ontogenesis of wheat variety Garagylchyg-2: A – sucrose (1), starch (2); B - the same as A (% of the total CO_2 fixation rate) (1, 2), rate of sucrose to starch (3); C – glycine-serine; D – the same as C (% of the total CO_2 fixation rate) (1), total value of glycine-serine (2).

Table data shows that the CO_2 assimilation rate in the earing stage averages 12.5 μmol $CO_2/dm^{2\cdot}$min in intense Garagylchyg-2 and Vugar genotypes, while in extensive Gyrmyzy bugda and Shark genotypes - 9.8 μmol $CO_2/dm^{2\cdot}$min.

At the same time the rate of malate, aspartate and alanine synthesis is 0.68, 0.51 and 0.42 μmol $CO_2/dm^{2\cdot}$min in intense genotypes at this stage and

0.59, 0.51 and 0.33 μmol $CO_2/dm^2 \cdot min$, respectively, in extensive genotypes. Apparently the rates of malate, aspartate and alanine labeling in the studied genotypes are similar in the milk ripeness stage as well.

Intensive and extensive genotypes do not strongly differ by the starch synthesis rate, which is 6.8% of the total assimilation rate. In the earing stage the glycine + serine biosynthesis rate in intense genotypes is 2.65 μmol $CO_2/dm^2 \cdot min$, while in extensive ones it is 1.62 μmol $CO_2/dm^2 \cdot min$, whereas in the milk ripeness stage it is 1.2 and 1.1 μmol $CO_2/dm^2 \cdot min$, respectively.

If we estimate the photorespiration by the glycine + serine synthesis rate, it is evident that it is greater in low-growing high-yielding intense varieties. Labeling of sucrose in the earing stage reaches an average of 10.6 in intense genotypes and up to 7.7 μmol $CO_2/dm^2 \cdot min$ in extensive ones, which account for 82% of the photosynthesis rate (in the milk ripeness stage - 78%). Comparison of the obtained data shows that in different wheat genotypes mainly the sucrose, glycolate pathway products - glycine + serine and, to a minor extent, amino acids and organic acids, such as malate, aspartate, alanine, are generated during ^{14}C metabolism.

At the same time, intense and extensive wheat genotypes do not virtually differ by the rate of starch, and also malate, aspartate and alanine labeling. Rate of biosynthesis of glycolate pathway products and intensity of CO_2 assimilation are higher in low-growing intense genotypes.

Thus, the rates of $^{14}CO_2$ incorporation into starch, alanine, malate, and aspartic acid in the intensive and extensive varieties of winter wheat were similar. The rates of ^{14}C incorporation into glycolate metabolites and sucrose, as well as $C0_2$ assimilation rates, were higher in the dwarf intensive varieties. As noted above, the high-yield varieties are characterized by high RuBP oxygenase activity and a high rate of photorespiratory CO_2 release.

These results demonstrate that despite increased photorespiration in the dwarf varieties, they show high rates of net photosynthesis due to increased true photosynthesis.

These varieties, produced by plant breeders, have optimal architectonics for the efficient use of solar energy. The high rate of net photosynthesis in these varieties is due to enhanced true photosynthesis provided by high activities of RuBP carboxylase and carbonic anhydrase; the latter enzyme contributes to the efficient functioning of the former in the carboxylation sites of the chloroplasts.

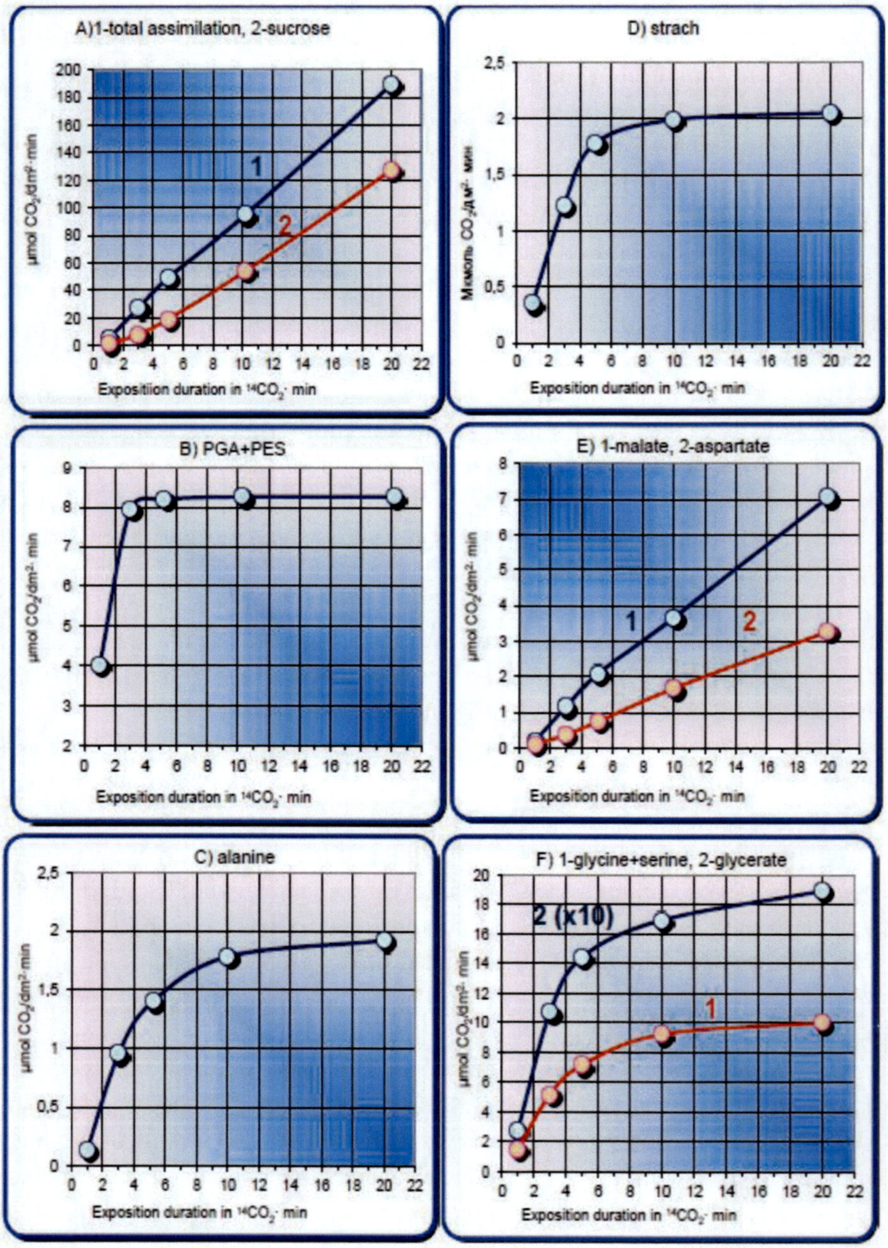

Figure 16. ^{14}C incorporation into photosynthesis products of flag leaf of Gyrmyzy bugda in the earing stage (μmol $CO_2/dm^{2}\cdot min$).

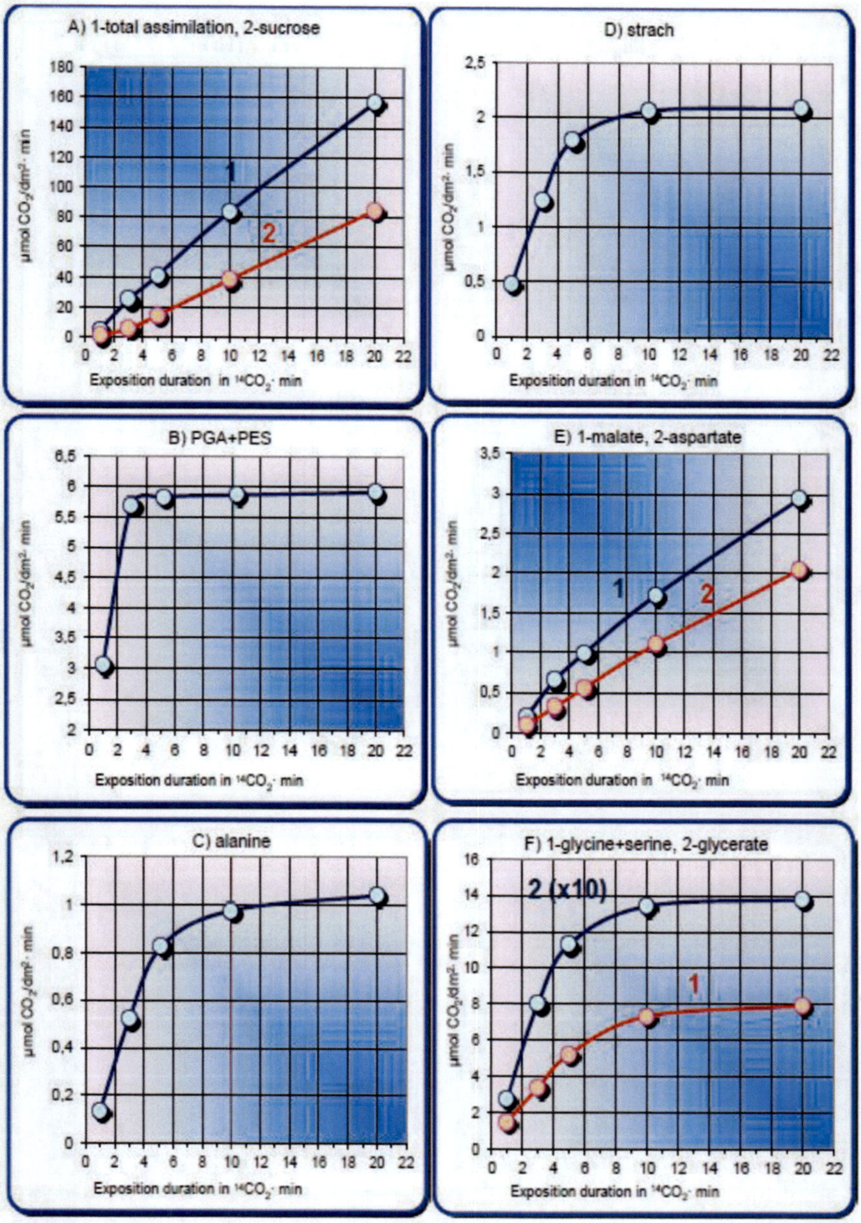

Figure 17. $^{14}CO_2$ incorporation into photosynthesis products of flag leaf Gyrmyzy bygda in the milk ripeness stage (μmol $CO_2/dm^2 \cdot min$).

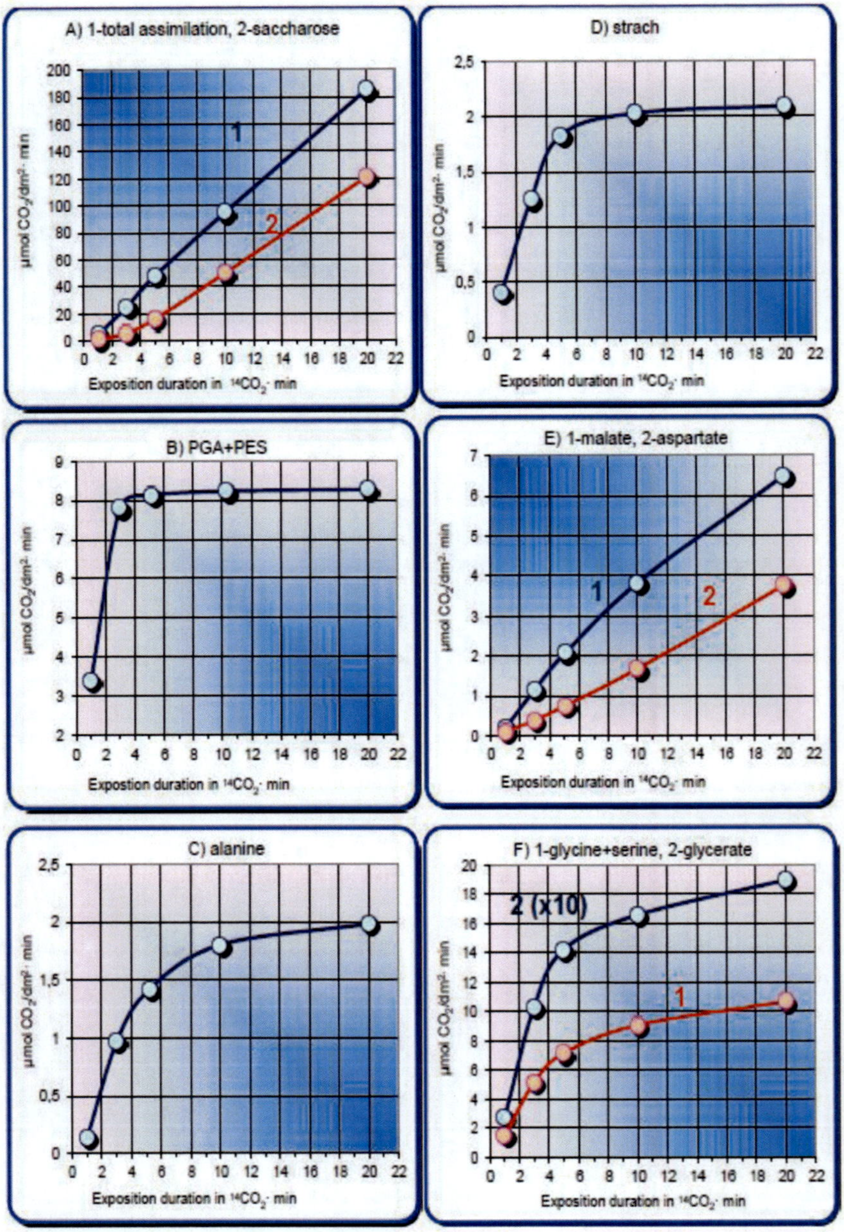

Figure 18. $^{14}CO_2$ incorporation into photosynthesis products of flag leaf of Sharg in the earing stage (μmol $CO_2/dm^2 \cdot min$).

Table 2. Kinetic characteristics of photosynthetic carbon metabolism in different varieties of winter wheat at the earing and milk ripeness stages

Genotypes		Short-stemmed, intensive				Long-stemmed, extensive			
		Garagylchyg-2		Vugar		Gyrmyzy bugda		Sharg	
Development stage Metabolites		Earing	Milk ripeness	Earing	Milk ripeness	Earing	Milk ripeness	Earing	Milk ripeness
Total assimilation (μmol CO_2/dm$^2\cdot$min)		12,3	7,0	12,9	6,9	9,6	7,1	9,9	7,5
Alanine	1	0.41	0,21	0,38	0,23	0.36	0,23	0,34	0,25
	2	3,3	3,0	3,1	2,6	3,8	3,2	4,0	3,6
Malate	1	0,70	0,29	0,64	0,27	0,59	0,32	0,61	0,34
	2	5,7	4,1	5,5	0,39	6,1	4,5	5,5	5,0
Aspartate	1	0,49	0,23	0,47	0,25	0,47	0,20	0,50	0,23
	2	4,0	3,3	4,2	3.5	4,9	2,8	5,6	3,1
Glycine+serine	1	2,5	1,1	2,9	1,4	1,57	1,0	1,6	1,2
	2	20,7	15,7	19,5	15,9	16,4	14,7	16,9	16,0
	3	20,4	11,6	21,6	10,3	11,5	8,4	12,4	9,0
Saccharose	1	10,6	5,5	11,3	6,1	7,8	5,5	8,5	6,1
	2	86,0	78,0	83,0	80,0	81,0	77	82	81
Starch	1	0,70	0,45	0,68	0,5	0,70	0,50	0,75	0,52
	2	5,7	6,4	5,2	7,0	7,3	7,0	7,1	8,1

1 – synthesis rate (μmol CO_2/dm$^2\cdot$min), 2 – the same figure (as % from total CO_2 fixation rate), 3 - total pool value (μmol CO_2/dm$^2\cdot$min).

It should be noted that the products of glycolate metabolism can also be used in sucrose synthesis or be transported from the leaves. Therefore, under certain conditions, the products of glycolate metabolism can contribute to active assimilate transport, thereby creating conditions for the maintenance of photosynthesis at a high level. In studies of plant productivity, sink source relationships between plant organs, which are the major endogenous factor regulating photosynthesis in the whole plant, attract increasing interest (Mokronosov, 1988). Hence, investigations of changes in photosynthetic carbon metabolism and related enzyme activities upon changing sink source relationships may be important in elucidating the mechanisms that underlie the Warburg effect and regulate the photosynthesis/photorespiration ratio (Chikov, 1987). Our data showed that disturbances in sink source relations in different wheat varieties led to changes in enzyme activities and the rates of photosynthate labeling (Figure 20-25). During the first day after the removal of the lower leaves, the flag leaf of the low-yield variety Gyrmyzy bygda showed a decline followed by an increase in the activities of the studied enzymes. After the removal of the lower leaves in the Gargylchyg-2 variety, the enzyme activities in flag leaves slightly increased initially and then returned to the control level (Figure 20-23). In the flag leaves of the low-yield variety Gyrmyzy bygda, ^{14}C incorporate into sucrose only slightly increased on the first day after leaf removal, while ^{14}C incorporation into glycine-serine decreased. During the following days, the rate of sucrose labeling increased, attaining a level higher than in the control plants. ^{14}C incorporation into serine-glycine also increased in the subsequent days to attain the control level, but ultimately declined to a level much lower than in the control plants (Figure 24-25). During the first days after the removal of the lower leaves in Garagylchyg-2 plants, sucrose and glycine-serine labeling in the flag leaves did not significantly differ from the control plants. Sucrose labeling remained unchanged during the following days, while it slightly decreased in the control plants. Glycine-serine labeling increased in the high-yield genotype, attained the control level, and then drastically decreased (Figure 24-25). The removal of a part of the spike in Gyrmyzy bygda wheat resulted in reduced activities of all the studied enzymes, including RuBP oxygenase; this was accompanied by lowered ^{14}C incorporation into glycine-serine. In contrast to Gyrmyzy bygda, a significant reduction in sucrose labeling was seen after spike removal in the high-yield variety Garagylchyg-2. Our results show that the rate of ^{14}C incorporation into the products of glycolate metabolism upon artificially disturbed sink-source relationships is consistent with the level of RuBP oxygenase activity. Opposite changes in the rates of sucrose and glycolate metabolite labeling were induced by altered sink source relationships both in high- and low-yield varieties. The results also show that each variety is characterized by a certain ratio of RuBP carboxylase to oxygenase activities, which could be only transiently changed by various treatments, specifically, by disturbing sink-source relationships between plant organs. Thus, we can suppose that the enzymes participating in the primary stages of CO_2 fixation are also involved in plant adaptation to variable environmental conditions.

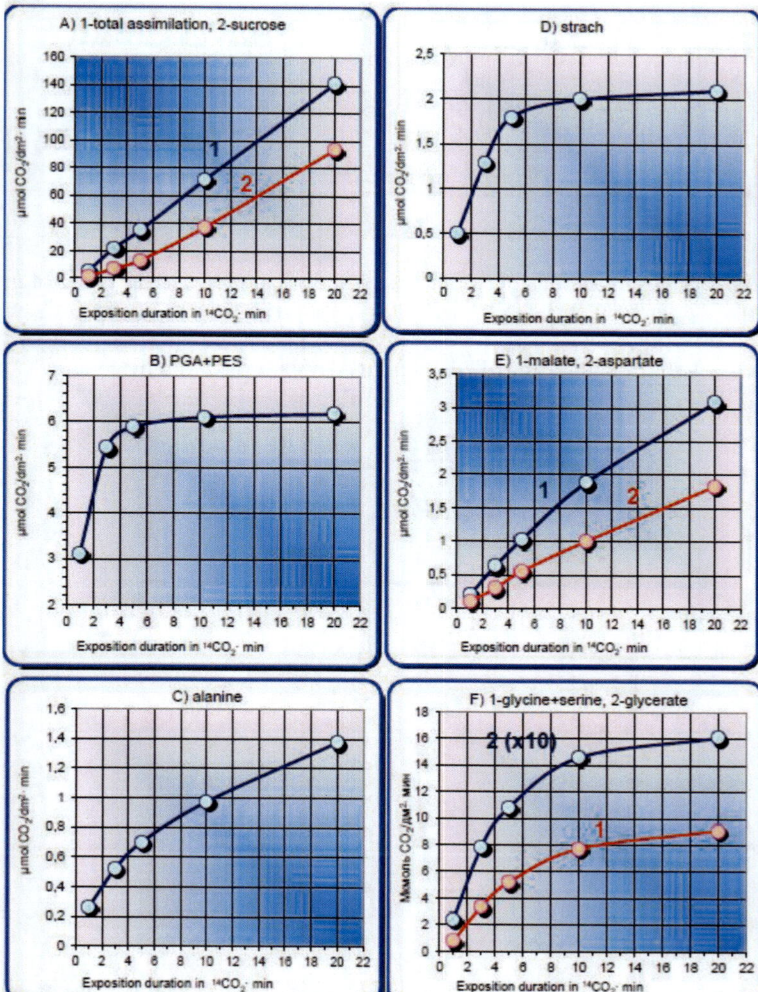

Figure 19. $^{14}CO_2$ incorporation into photosynthesis products of flag leaf of Sharg in the milk ripeness stage (μmol CO_2/dm$^{2}\cdot$min).

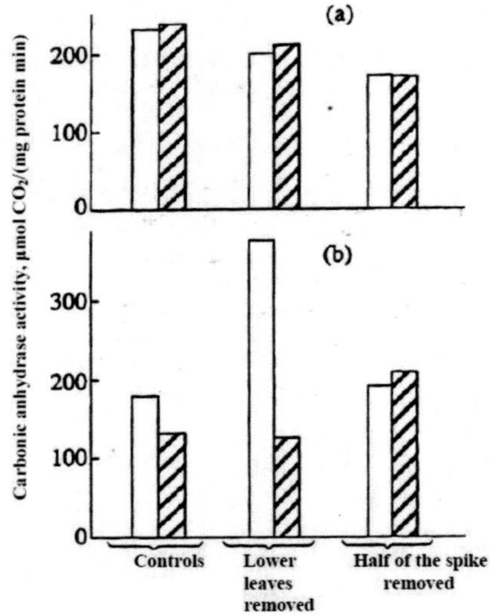

Figure 20. Changes in carbonic anhydrase activity in flag leaves in the phase of grain filling after the removal of the lower leaves or half of the spike in different wheat varieties. (a) Gyrmyzy bygda; (b) Garagylchyg-2. Carbonic anhydrase activity on the first (open columns) and fourth (filled columns) days after leaf removal.

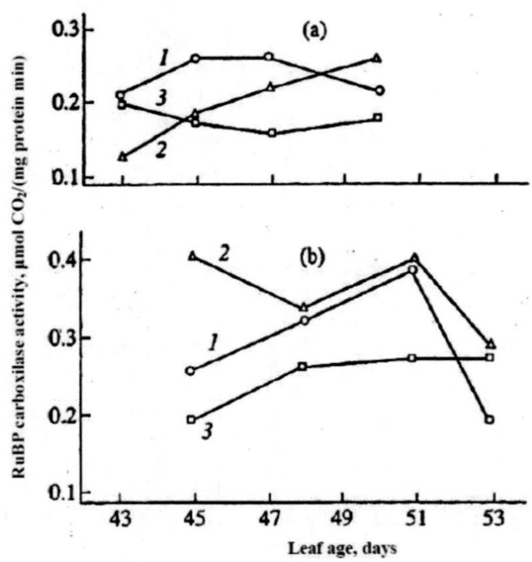

Figure 21. Changes in RuBP carboxylase activity in flag leaves in the phase of grain filling after the removal of the lower leaves or half of the spike in different wheat varieties. (a) Gyrmyzy bygda; (b) Garagylchyg-2. (1) Control plants; (2) lower leaves removed; (3) half of the spike removed.

The adaptation processes in different plant varieties have both common and distinctive features associated with the plant genotype. In conclusion, our data on the characteristics of photosynthetic carbon metabolism in different varieties of winter wheat contribute to understanding photosynthetic mechanisms, as such, and the mechanisms providing high photosynthetic productivity; these data can also be used by plant breeders in developing new high-yield wheat varieties.

Table 3 represents data on photosynthate distribution between plant organs in the flowering phase 24 h after labeling. It is evident that assimilate export to different plant organs is genotypically determined and depends on the location of the treated leaf. When the label was introduced into the flag leaf, the percentage of the label found in the top internode was substantially higher in the dwarf (Garagylchyg-2) variety than in the tall varieties.

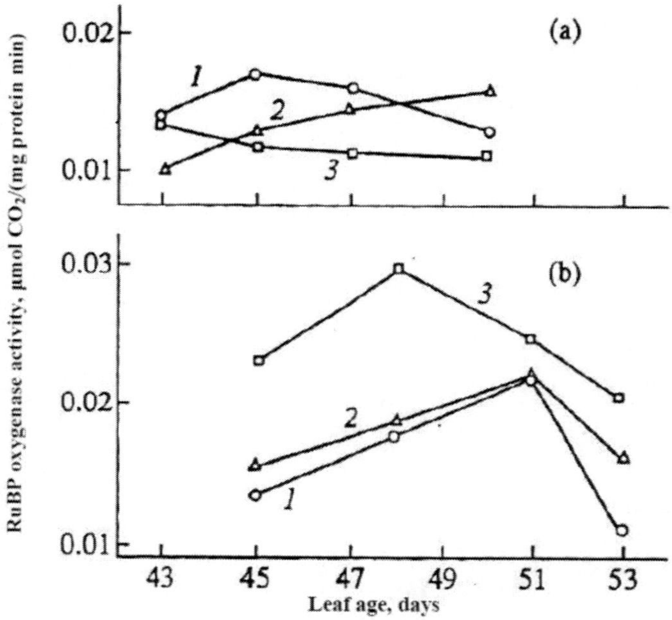

Figure 22. Changes in RuBP oxygenase activity in flag leaves in the phase of grain filling after the removal of the lower leaves or half of the spike in different wheat varieties. (a) Gyrmyzy bygda; (b) Garagylchyg-2. (1) Control plants; (2) lower leaves removed; (3) half of the spike removed.

In Garagylchyg-2, the top internode contained about 26% of the total radioactivity, whereas only 12-14% of the total radioactivity was found in the top internode of tall varieties.

The dwarf and tall varieties also markedly differed in the label content in lower internodes. In Garagylchyg-2, lower internodes contained 10.3% of the total radioactivity, whereas the lower internodes of tall varieties accumulated 27% (Gyrmyzy bygda) or 40% (Kansas-63323) of the total radioactivity.

At the flowering phase, a significant portion of assimilates was exported from flag leaves to ears in the dwarf variety (about 22%), whereas only 5-7% of the assimilates were exported to ears in the tall varieties.

Similar differences between tall and dwarf varieties in the pattern of assimilate export were found in experiments with labeled lower leaves. We found no marked difference between the varieties in the rate of assimilate export from flag leaves (as judged from the residual radioactivity of flag leaves), although significant differences were found when the label was introduced into the sixth or seventh leaf from the base.

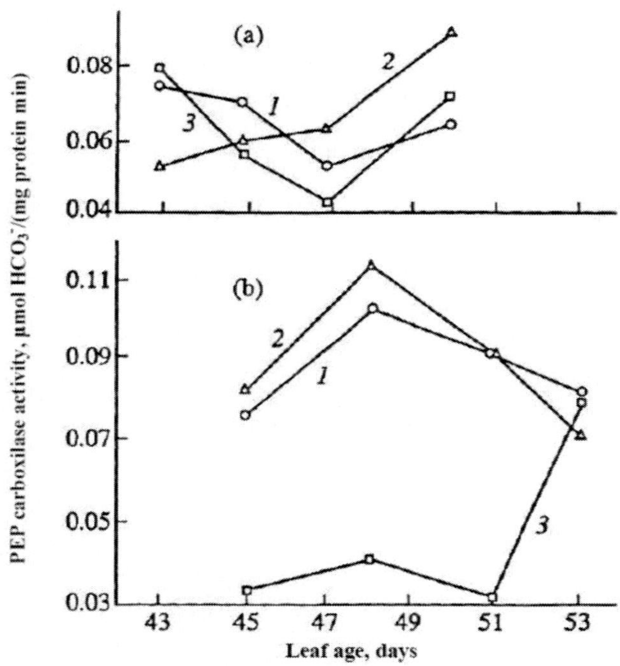

Figure 23. Changes in PEP carboxylase activity in flag leaves in the phase of grain filling after the removal of the lower leaves or half of the spike in different wheat varieties. (a) Gyrmyzy bygda; (b) Garagylchyg-2. (1) Control plants; (2) lower leaves removed; (3) half of the spike removed.

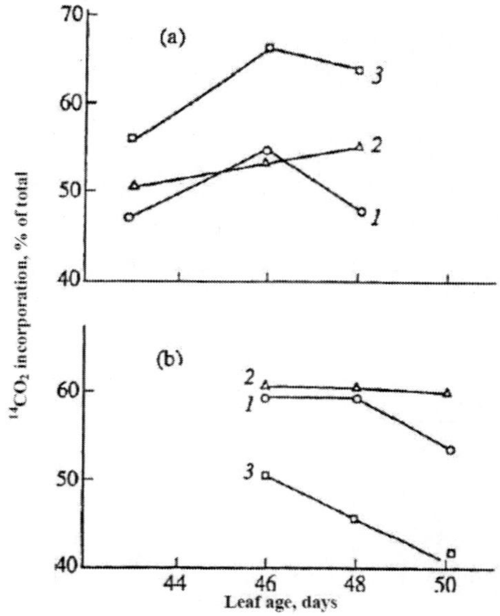

Figure 24. [14]C incorporation into sucrose in flag leaves in the phase of grain filling after the removal of the lower leaves or half of the spike in (a) Gyrmyzy bygda (b) Garagylchyg-2. (1) Control plants; (2) lower leaves removed; (3) half of the spike removed.

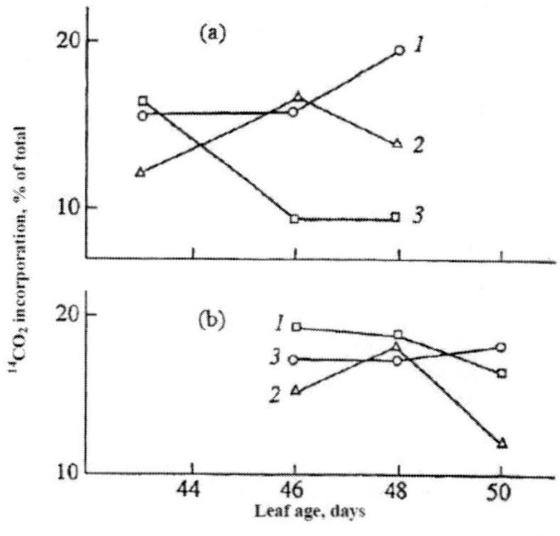

Figure 25. [14]C incorporation into sucrose in flag leaves in the phase of grain filling after the removal of the lower leaves or half of the spike in (a) Gyrmyzy bygda (b) Garagylchyg-2. (1) Control plants; (2) lower leaves removed; (3) half of the spike removed.

Table 3. Assimilate distribution between organs of different wheat genotypes within 24 h after plant feeding with $^{14}CO_2$ during the flowering phase (% of total plant radioactivity)

Organ	Gyrmyzy bygda			Garagylchyg-2			Kansas-63323
	Storey of ^{14}C-source leaf						
	8 (flag leaf)	7	6	8	7	6	8
Fed leaf	39.4±1.4	61.9±3.4	73.4±5.4	30.7±2.1	46.3±3.7	60.9±4.4	32.1±2.0
Ear	7.4±0,4	3.7±0,2	2.0±0.1	22.5±1.3	8.6±0.9	5.3±0.4	5.5±0.4
Stem above the source leaf	12.4±0.5	7.9±0.3	6.1±0.3	26.5±1.2	16.5±0.8	12.0±0.9	13.6±0.6
Stem below the source leaf	27.1±0.8	13.7±0.5	9.2±0.7	10.3±0.6	20.2±1.1	13.8±1.4	40.3±1.9
Sheath of the source leaf	11.3±0.7	10.3±1.1	6.3±0.4	9.0±0.6	6.6±0.3	5.6±0.3	6.6±0.5
Root	2.4±0.2	2.5±0.2	3.0±0.2	1.0±0.1	1.8±0.2	2.4±0.2	2.8±0.3

Table 4. Assimilate distribution between organs of different wheat genotypes within 24 h after plant feeding with $^{14}CO_2$ at the milky ripeness phase (% of total plant radioactivity)

Organ	Gyrmyzy bygda		Garagylchyg-2		Kansas-63323
	Storey of ^{14}C-source leaf				
	8 (flag leaf)	7	8	7	8
Fed leaf	27.6±2.0	33.4±1.9	26.7±1.3	30.1±1.4	29.2±1.9
Ear	52.4±3,8	18.9±0,9	66.1±4.0	55.0±3.4	57.9±3.8
Stem above the source leaf	9.3±0.4	13.1±1.0	2.1±0.2	7.1±0.5	2.1±0.8
Stem below the source leaf	5.5±0.2	23.9±3.2	1.7±0.1	3.7±0.2	6.1±0.4
Sheath of the source leaf	4.5±0.3	7.2±0.8	2.6±0.2	3.6±0.2	2.7±0.6
Root	0.7±0.06	3.5±0.2	0.8±0.06	0.5±0.04	2.0±0.1

Table 5. Assimilate distribution between organs of different wheat genotypes at the end of growth period. $^{14}CO_2$ was fed at the milky ripeness phase (% of total plant radioactivity)

Organ	Gyrmyzy bygda		Garagylchyg-2		Kansas-63323
	Storey of ^{14}C-source leaf				
	8 (flag leaf)	7	8	7	8
Fed leaf	18.1±0.8	29.1±2.3	17.5±0.9	18.1±1.1	18.5±0.9
Ear	78.2±3,9	55.3±3,1	81.7±4.4	80.3±4.8	79.0±4.2
Stem above the source leaf	0.6±0.04	5.8±0.8	0.2±0.01	0.6±0.04	0.5±0.06
Stem below the source leaf	0.9±0.06	3.6±0.2	0.3±0.02	0.2±0.02	0.9±0.07
Sheath of the source leaf	2.0±0.09	5.3±0.4	0.3±0.03	0.8±0.08	1.1±0.07
Root	0.7±0.06	3.5±0.2	0.8±0.06	0.5±0.04	2.0±0.1

Tables 3 and 4 compare the export and distribution of assimilates at the flowering and milky ripeness phases. The following common features were observed in all varieties studied:

(1) For all leaves, the rates of assimilate export were higher at the milky ripeness phase than at the flowering phase.
(2) At the phase of milky ripeness, assimilate export from leaves and assimilate import into ears were most rapid during the first day after labeling. During this day, flag leaves exported up to 70% of the labeled assimilates to the ears. During the subsequent period until complete ripeness, no more than 10-20% of assimilates were exported from leaves to ears, and 10-30% of the initial radioactivity remained in the treated leaves by the end of the growth period (Table 3 and 4).
(3) In contrast to the label infusion at the flowering phase, assimilate labeling in the milky phase resulted in a significant increase in ^{14}C content in the ears of all varieties. Correspondingly, the accumulation of radioactivity in other plant organs was reduced, indicating a higher sink activity of the ears during the milky ripeness phase.

Significant genotypic differences in assimilate accumulation in ears were found during the milky ripeness phase; assimilates accumulated more actively in the ears of dwarf varieties than in tall ones (Table 4 and 6).

Furthermore, at the milky ripeness phase, more assimilates were exported to ears from the seventh leaf in the dwarf plants than in tall genotypes, indicating a higher contribution of the seventh leaf to grain filling in the dwarf variety. As can be seen from Tables 3 and 6, assimilates formed in the flowering phase were subject to radical redistribution during the period from flowering to complete grain ripeness. During this period, the ^{14}C content in the ears significantly increased at the expense of assimilates remobilized from the stem and leaves.

For example, by the time of complete grain ripeness, about 15% of the radioactivity remained in the labeled flag leaves, whereas a substantial part of ^{14}C (up to 25%) was found in the stems. However, the amount of 14C exported from leaves to ears over the period from flowering to grain ripeness was 10-20% higher in the dwarf than in the tall varieties. The difference between these varieties was also characteristic of ^{14}C content in the stems; less assimilates were retained in stems of the dwarf variety than in the stems of tall varieties.

At the phase of complete milky ripeness (Table 5), assimilates were almost fully exported to the ears without being retained in the stems. In all varieties studied, about 80% of ^{14}C-labeled assimilates were exported from the leaves to the ears.

Thus, our investigation of the assimilate transport and distribution in different varieties of winter wheat has shown that assimilates formed at the early phases of reproductive development are more efficiently used for grain formation and, to a lesser extent, spent for vegetative biomass formation in the dwarf variety than in the tall (extensive) varieties. Moreover, lower leaves of the short-stem variety are more active in grain filling than the lower leaves of tall varieties.

Our data on the variety-specific features of the export and distribution of assimilates between plant organs contribute to the understanding of the mechanisms that underlie high yields in plants. Our results supplement the data obtained in other physiological and biochemical studies (Aliev, 1983; Kazibekova et al., 1985). These data can be used by breeders to obtain high-yielding cultivars of wheat or other crops.

The final result of the plant development, all the complex physiological-biochemical processes, occurring in different plant organs during ontogenesis is grain yield and quality, providing the growth of the plant vegetative mass as well as accumulation of the storage compounds in reproductive organs. At the same time yield quality in wheat and other crops, first of all, is related to the protein content of grains, where not only total protein content but also quality of the protein complex is of great importance, since the food value of protein is specified by the content of indispensable amino-acids.

Characteristics of metabolism, including proteins in vegetative organs have a great effect on yield and protein content in grains. Thereby the study of the protein exchange features in vegetative organs at different donor-acceptor relationships is very important.

However, the study of protein exchange only in wheat grains is insufficient for understanding the regularity of the protein quality formation in grains and elucidation of the interactions of its vegetative and reproductive organs. Considering this circumstance, we carried out experiments to determine the protein content in different parts of the vegetative organs in several wheat genotypes at different stages of ontogenesis: earing, flowering and milk ripeness phases.

Table 7 and 8 present data on the protein content of different fractions at separate stages of ontogenesis of the genotypes-Garagylchyg-2 and Gyrmyzy bugda. Investigations of the protein exchange in leaves and stem show that in both genotypes protein content in relative units is higher in leaves compared with the stem and it decreases during ontogenesis (Table 7).

Table 6. Assimilate distribution between organs of different wheat genotypes at the end of growth period. $^{14}CO_2$ was fed at the flowering phase (% of total plant radioactivity)

Organ	Gyrmyzy bygda			Garagylchyg-2			Kansas-63323
	Storey of ^{14}C-source leaf						
	8 (flag leaf)	7	6	8	7	6	8
Fed leaf	13.6±1.1	17.6±1.0	31.1±2.7	12.8±0.8	16.4±1.1	25.9±1.9	20.5±1.8
Ear	55.2±4,2	42.0±3,1	29.6±1.7	74.5±4.9	67.9±4.3	56.1±3.3	56.4±2.9
Stem above the source leaf	13.6±0.8	18.1±0.9	21.8±1.2	5.3±0.4	5.3±0.8	9.3±1.0	5.5±0.6
Stem below the source leaf	11.5±0.5	14.9±0.6	12.1±0.7	4.2±0.3	6.5±0.4	4.8±0.4	12.1±0.5
Sheath of the source leaf	5.0±0.4	6.1±0.4	4.0±0.3	2.5±0.08	2.5±0.2	2.1±0.1	3.5±0.4
Root	1.2±0.06	1.3±0.1	1.4±0.09	0.7±0.06	1.4±0.1	1.8±0.1	2.0±0.2

Table 7. Comparison of the protein contents of different fractions in leaves and stems of wheat (imp/min per 1g of the dry weight)

Garagylchyg-2								
Fractions	Water-soluble		Salt-soluble		Alcohol-soluble		ALKALI-SOLUBLE	
Phases	Leaves	Stem	Leaves	Stem	Leaves	Stem	Leaves	Stem
Earing	720	378	99	52	115	109	387	316
Flowering	567	155	135	56	43	103	548	513
Milk ripeness	657	180	104	43	105	95	277	204
Qyrmyzy bugda								
Earing	790	424	108	60	130	131	409	346
Flowering	611	170	160	63	49	110	606	565
Milk ripeness	717	185	117	45	120	98	295	207

Table 8. Total content of soluble proteins in leaves and stems of wheat during ontogenesis (imp/min per 1g of the dry weight)

Garagylchyg-2			
Phases / Organs	Earing	Flowering	Milk ripeness
Leaves	1321	1293	1143
Stem	926	827	522
Gyrmyzy bugda			
Leaves	1437	1366	1250
Stem	961	908	535

The decline in the total and protein nitrogen per unit of weight in leaves and other organs with aging is already well known. Water- and alkali-soluble proteins, which contribute most to the total value of the soluble proteins, prevail in vegetative organs (Table 8). It is obvious that, for Garagylchyg-2 water-soluble protein contents in leaves are 720 imp, 567 imp and 657 imp per gram of the dry weight during the phases of earing, flowering and milk ripeness, respectively. While in the stem these values are obtained to be 378, 155 and 180 imp/min per gram of the dry weight i.e. they decrease during the flowering phase, then slightly increase during the phase of the milk ripeness but remain less compared with the earing phase. The comparison of the obtained data also shows that the contents of the water- and salt-soluble proteins are more in leaves than in stem. Due to the low contents of the water-soluble proteins in stem compared with leaves a large portion of alkali- and alcohol-soluble proteins i.e. the storage proteins was observed in the stem protein complex, contrary to leaves where water-soluble proteins prevailed during all developmental phases. It is well known in the scientific literature that stem functions as a temporary storage of reserve compounds, including proteins, which subsequently used during grain filling period and albumins dominate in the protein complex of young leaves and stems of wheat that is confirmed by our data. Analogical pattern has been found for the protein allocation in different fractions of vegetative organs of Gyrmyzy bugda. However, in all phases the protein content appeared to be slightly higher, than in Garagylchyg-2, apparently it is related to the fact that Gyrmyzy bugda is a high-protein genotype. It should be noted that the protein content in different fractions of wheat vegetative organs changes with age, though not to the same extent. Relative contents of water-soluble proteins in leaves and stem sharply decrease during the flowering phase compared with the earing phase. Similar changes occur in the content of alcohol-soluble proteins in leaves and a slight decrease is observed in the stem with age.

Alkaline-soluble proteins accumulate in vegetative organs before the flowering phase and then their content decreases. Changes in the contents of salt- and alkali-soluble proteins in leaves occur by the unimodal curve type. It is more pronounced for stem alkali-soluble proteins.

Table 9. Wheat grain protein content when changed sink source relations

Garagylchyg-2			
Variants	Total protein content		Protein content in one ear, g
	imp/min per 1g of dry weight	%	
Control	1096	13,7	0,278
REMOVED 8TH LEAF	1008	12,6	0,208
removed 7th leaf	1040	13,0	0,214
removed ½ ear	1144	14,3	0,130
Gyrmyzy bugda			
CONTROL	1208	15,1	0,231
removed 8th leaf	1128	14,1	0,183
removed 7th leaf	1144	14,3	0,197
removed ½ ear	1248	15,6	0,114

The comparison of the changes in the total soluble protein content in leaves and stems of different wheat genotypes during ontogenesis is of great interest (Table 8). It is obvious that during the transition from the earing phase to the phase of milk ripeness the total protein content in leaves of Garagylchyg-2 decreases from 1321 to 1143 imp/min, which is 14%, while in the stem a decrease from 926 to 522 imp/min occurs, which is about 44%. In Gyrmyzy bugda these parameters are 13 and 44.3%, respectively. Though the relative decrease of the soluble protein content during the transition from the earing phase to the phase of milk ripeness in vegetative organs of different wheat genotypes differs slightly, in Gyrmyzy bugda absolute values of these parameters are higher.

According to modern views, production process in wheat and other crops mostly depends on the interaction between assimilating and consuming organs.

At the same time, wheat grain quality, including protein content, largely depends on the state of the donor-acceptor relations in the whole plant system.

Table 9 shows the analysis of the quality of grain yield in different wheat genotypes-Garagylchyg-2 and Gyrmyzy bugda, which donor-acceptor relations have been changed by removing flag and sub flag leaves during phases of earing, flowering and by removing half of the ear during the phase of milk ripeness. When removing the 8th leaf the 7th leaf was labeled with $^{14}CO_2$ and when removing the 7th leaf the label was introduced into the 8th leaf and after removing half of the ear the labeling was performed with both leaves.

It should be noted that the data obtained with the same wheat genotypes representing protein percentage contents in grains and protein contents in one ear in grams, according to the protein nitrogen, are most consistent with the results obtained with the $^{14}CO_2$ use (imp/min per 1g of the dry weight). The comparison of these data shows that when removing the 8th, 7th leaves and half of the ear, the protein content in grains of Garagylchyg-2 is 1008 imp/min (12.6%), 1040 (13.0%) and 1142 imp/min per 1g of the dry weight (14.3%), respectively compared with the control variant-1096 imp/min (13.7%) (Table 3).

Whereas for the genotype Gyrmyzy bugda these parameters are 1128 imp/min (14.1%), 1144 imp/min (14.3%) and 1248 imp/min (15.6%), respectively, compared with the control-1208 imp/min (15.1%). It is evident that the remove of the flag and sub flag leaves in both genotypes leads to a decrease of the protein content in grains in one ear compared with the control plants, since the remove of the 8th and 7th leaves causes a change in the ratio between the nitrogen amount in vegetative organs and the plant grain mass. Therefore the outflow of nitrogen compounds from vegetative organs to grains decreases and the decrease in the grain protein content of both genotypes is more pronounced when removing the 8th leaves than the 7th leaves. There is a positive correlation between contents of the grain protein and leaf nitrogen and the highest correlation is observed between contents of the grain protein and flag leaf nitrogen. Our results confirm that the absence of the 8th leaf is more appreciable in the process of the protein accumulation in grains than the absence of the 7th leaf.

The increase in the protein content when removing half of the ear may be elucidated by the fact that decreases in the grain amounts of ear leads to a change of the ratio between the grain weight and the plant vegetative mass, when the same amounts of nitrogen compounds in vegetative organs match the fewer grains compared with the control variant.

It is noteworthy that the protein content in one ear of both genotypes decreases when removing half of the ear, which can be explained by approximately a 2-fold decrease of the ear grain mass.

The changes of the donor-acceptor relations between assimilating and consuming organs of different wheat genotypes in all cases lead to the protein content changes in the grain yield and in one ear.

So in conclusion, the high intensity of true photosynthesis and photorespiration, high activity of primary photochemical processes together with favorable phonotypical traits, an optimum leaf area index and best architectonics define high productivity of wheat genotypes. Our data on the variety-specific features of the export and distribution of assimilates between plant organs contribute to the understanding of the mechanisms that underlie high yields in plants. Our results supplement the data obtained in other physiological and biochemical studies (Aliev, 1983, Kazibekova et al., 1985). These data can be used by breeders to obtain high-yielding cultivars of wheat or other crops.

REFERENCES

Akhmedov, G.A. (1986) CO_2 exchange in wheat ontogenesis depending on phenotypic traits, growth conditions and photosynthetic productivity. PhD thesis, Baku: 183 p. (in Russian).

Aliyev, J.A. (1974) Photosynthetic activity, mineral nutrition and plant productivity. Baku, Elm: 335 p. (in Russian).

Aliyev, J.A., Kazibekova, E.G. (1977) About architectonics and photosynthetic function of high-yielding wheat. *Russian J. Plant Physiol.,* 24(5): 662-667 (in Russian).

Aliev, D.A., Azizov, I.V. (1977) Photochemical activity of winter wheat chloroplasts in relation to their optical and biological structure, productivity and irradiance. Izv. Akad. Nauk Azerb. SSR, (ser. biol. nauk), 5: 69-74 (in Russian).

Aliev, D.A., Kazibekova, E.G. (1979a) Peculiarities of photosynthetic rate in extensive and intensive wheat cultivars. *Izv. Akad. Nauk Azerb. SSR,* (ser. biol. nauk), 3: 10-15 (in Russian).

Aliev, D.A., Kazibekova, E.G. (1979b) Photosynthesis of extensive and intensive wheat varieties. Izv. Akad. Nauk Azerb. SSR, 3: 3-10 (in Russian).

Aliyev, J.A., Kazibekova, E.G., Safarov, S.A. (1982) The main parameters of leaves of wheat phenotypes with various yield. Izv. Akad. Nauk Azerb. SSR, (ser. biol. nauk), 5: 3-8 (in Russian).

Aliyev, J.A. (1983) Modern notion of the perfect wheat plant. Izv. Akad. Nauk Azerb. SSR, (ser. biol. nauk) 3: 3-14 (in Russian).

Aliev, D.A., Kazibekova, E.G. (1988) Genetic analysis of the characteristics of the photosynthetic apparatus in relation to winter wheat productivity. Fotosintez i produktsionnyi protsess (Photosynthesis and Productivity), Moscow: Nauka, 237 p.

Aliev, J.A., Guliev, N.M., Kerimov, S.Kh., Hidayatov, R.B. (1988) Enzymes of the primary CO2 fixation in flag leaf ontogenesis of wheat genotypes. Izv. Akad. Nauk Azerb. SSR, (ser. biol. nauk), 4: 12-20 (in Russian).

Aliev, D.A., Guliev, N.M., Idayatov, R.B. (1989) Isolation and some properties of carbonic anhydrase obtained from wheat leaves. Izv. Akad. Nauk Azerb. SSR, (ser. biol. nauk), 5: 15-29 (in Russian).

Aliyev, D.A., Akhmedov, A.A., Mirzoyev, R.S. (1992) The dynamics of the components of carbon dioxide gas exchange of soybean leaves in the field condition. Izv. AN Azerb. SSR, (ser. biol. nauk), 1-6: 76-83 (in Russian).

Aliev, J.A., Kazibekova, E.G. (1995) Peculiarities of highly productive wheat photosynthesis and usage of photosynthetic signs in selection. Proceedings of the Xth International Congress on Photosynthesis, Montpellier, France, 1995. In: Photosynthesis: from light to Biosphere (Mathis P., ed.), Kluwer Academic Publishers, Dordrecht, Boston, London, 54: 659-662.

Aliev, D.A., Kerimov, S.Kh., Dzhangirov, A.A., Akhmedov, A.A. (1996 a) Transport and distribution of 14C-assimilates in wheat genotypes with various photosynthetic characteristics and economic yields. *Russian J. Plant Physiol.,* 43(1): 57-61.

Aliev, D.A., Kerimov, S.Kh., Guliev, N.M., Akhmedov, A.A. (1996 b) Carbon metabolism in wheat genotypes with contrasting photosynthetic characteristics. *Russian J. Plant Physiol.,* 43(1): 42-48.

Aliev, J.A., Guliev, N.M., Kerimov, S.Kh., Hidayatov, R.B. (1996 c) Photosynthetic enzymes of wheat genotypes differing in productivity. *Photosynthetica,* 32(1): 77-85.

Aliyev, J.A. (2001) Diversity of photosynthetic activity of organs of wheat genotypes and breeding of high-yielding varieties tolerant to water stress. Proceedings of the 12th International Congress on Photosynthesis, Brisbane, Australia, www.publish.csiro.au/ps2001, S28-006.

Aliyev, J.A., Kazibekova, E.G. (2002) Peculiarities of high productive wheat photosynthesis and usage of photosynthetic signs in selection. Proceedings of Azerbaijan National Academy of Sciences (biological sciences), 1-6: 20-29 (in Russian).

Aliyev, J.A. (2006) Wheat breeding in Azerbaijan. Proceedings of Azerbaijan National Academy of Sciences (biological sciences), 3-4: 3-32 (in Russian).

Aliyev, J.A. (2011) Photosynthesis, photorespiration and productivity of wheat and soybean genotypes. Proceedings of Azerbaijan National Academy of Sciences (Biological Sciences), 66(1): 5-52.

Aliyev, J.A. (2012) Photosynthesis, photorespiration and productivity of wheat and soybean genotypes. *Physiol. Plant.,* 145(3):369-83

Bird, I.F., Cornelius, M.J., Keys, A.J. (1980) Effect of carbonic anhydrase on the activity of ribulose bisphosphate carboxylase. *J. Exp. Bot.,* 31: 365-369.

Bonnemain, J.L. (1982) Le transport des produits de la photosynthese lors development des plantes. C.R.Seances Acad. Agric. Fr., 68: 893.

Boyer, J.S. (1995) Measuring the water status of plants and soils. Academic Press, San Diego.

Bykov, O.D., Zelensicii, M.I. (1982) Improvement of photosynthetic performance by breeding crops. In: Fiziologiya fotosinteza (Physiology of Photosynthesis), Moscow: Nauka, p. 294 (in Russian).

Bykov, O.D., Zelenskii', M.I. (1987) Ecological and species wheat differentiation according to the level of photosynthetic activity. Sborn. Nauchn. Trudov Prikl. Bot. Genet. Selek. 100: 186-203 (in Russian).

Catalogue of varieties, released at Agriculture Institute of Azerbaijan (2000).

Chikov, V.I. (1987) Causes of variability in the magnitude of the Warburg effect among leaves differing in metabolite export, Fotosintez i transport assimilyatov (Photosynthesis and Assimilate Transport), Moscow: Nauka, 37 p.

Cook, M.G., Evans, L.T. (1978). Effect of relative size and distance of competing sinks on the distribution of photosynthetic assimilates in wheat. *Austral. J. Plant Physiol.,* 5(4): 495-509.

Dospekhov, B.A. (1985) Methods of field experience. Agropromizdat, Moscow: 351 p. (in Russian).

Evans, L.T., Dunstone, R.L. (1970) Some physiological aspects of evolution in wheat. - *Austr. J. biol. Sci.,* 23: 725-741.

Firus, O.K. (1986) Physiology of Carbon Assimilation by the Cyanobacterium Synechococcus cedrorum Sauv. Thesis. Dushanbe (in Russian).

Gavrilenko, V.F., Rubin, B.A., Zhigalova, T.V. (1974) Photosynthetic phosphorylation of isolated chloroplasts and photosynthetic rate in wheat seedlings of different productivity. *Sel'skokhoz. Biol.,* 9: 345-351 (in Russian).

Guliev, N.M., Kerimov, S.Kh., Jahangirov, A.A., Idayatov, R.B., Aliev, D.A. (1985) Activity of photosynthetic enzymes and CO2 assimilation in wheat genotypes. Tez. Dokl. Vsesoyuz. Simp. "Elementy Gazoobmena Lista i Tselogo Rasteniya i ikh Izmeneniya v Ontogeneze". USSR: Puschino, 15-16 (in Russian).

Jahangirov, A.A. (1987) Products of photosynthesis and peculiarities of their utilization with plants of different productivity. PhD thesis, Baku: 135 p. (in Russian).

Kalinkina, L.G., Rusinova, N.G., Strogonov, B.P. (1981) Effect of NaCl on oxygenase activity of ribulose- disphosphate carboxylase; activity of glycollate dehydrogenase and evolution of glycolic acid in the process of growth of sea and freshwater Chlorella. *Fiziol. Rast.,* 28: 5-13 (in Russian).

Kaplan, B.G. (1970) Express-calculation of the basic mathematical and statistical indicators. Maarif, Baku: 447 p. (in Russian).

Karpushkin, L.T. (1971) The use of infrared gas analyser to study CO2 gas exchange in plants. In: Biophysical methods in plant physiology (Molotkovskiy Yu.G., ed.). Nauka, Moscow: 44-71 (in Russian).

Kazibekova, E.G., Azizov, I.V., Aliev, D.A. (1985) Photosynthetic activity of chloroplasts in wheat cultivars of different productivity. Izv. Akad. Nauk Azerb. SSR, (ser. biol. nauk), 5: 3-10 (in Russian).

Keerberg, O.F., Vark, E.Y., Keerberg, Kh.I., Parnik, T.R. (1970) The study of kinetics of 14C incorporation in photosynthetic products in legume leaves. *DAN SSSR,* 195(1): 238-241 (in Russian).

Kerimov, S.Kh., Guliev, N.M., Idayatov, R.B., Jahangirov, A.A. (1987) Photochemical metabolism in different wheat genotypes. In: Trudy Vsesoyuz. mezhuniv. Konf. "Biologiya Kletki", pp. 210-211, Tbilisi (in Russian).

Khramova, G.A., Nizovskaya, N.V., Mamedov, N.D., Krendeleva, T.E., Paschenko, V.Z., Aliev, D.A. (1988) Primary processes of photosynthesis in chloroplasts of different varieties of hard wheat. Izv. Akad. Nauk Azerb. SSR, (ser. biol. nauk), 1: 3-9 (in Russian).

Kumakov, V.A. (1977) Selection aimed at increasing photosynthetic plant productivity. Itogi Nauki Tekhniki (ser. Fiziol. Rast.), 3: 108-125 (in Russian).

Kursanov, A.L. (1984) Endogenous regulation of assimilate transport and sink-source relations in plants. Fiziol. Rast. (Moscow), 31(3): 579, (Soy. Plant Physiol., engl. transl.).

Lloyd, N.D.H., Canvin, D.T. (1977) Photosynthesis and photorespiration in sunflower selections. *Can. J. Bot.,* 55: 3006-3012.

Lowry, O.H., Rosebrough, N.J., Farr, A.L., Randall, R.J. (1951) Protein measurement with the folin phenol reagent. *J. Biol. Chem.,* 193: 265-275.

Majeau, N., Coleman, J.R. (1994) Correlation of carbonic anhydrase and ribulose-l,5-bisphosphate carboxylase/oxygenase expression in pea. *Plant Physiol.,* 104: 1393-1399.

Methodological guidelines of Vavilov N.I. Institute of Plant Breeding (1987) Comprehensive physiological evaluation of drought- and heat-tolerant wheat in Uzbekistan (Kojushko N.N., ed). Leningrad: 23 p. (in Russian).

Milthorp, F.L. (1978) Approaches to yield physiology. *Fiziol. Biokhim. kul't. Rast.,* 339-349 (in Russian).

Miziorko, H.M. (1979) Ribulose-l,5-bisphosphate carboxylase. Evidence in support of the existence of distinct CO2 activator and CO2 substrate sites. *J. Biol. Chem.* 254: 270-272.

Mokronosov, A.T. (1978) Endogenous regulation of photosynthesis in the whole plant. Plant Fiziol (Moscow), 25(5): 938 (in Russian).

Mokronosov, A.T. (1981) Ontogeneticheskii aspect fotosinteza (Ontogenetic Aspect of Fotosyntesis). Moskow: Nauka (in Russian).

Mokronosov, A.T. (1988) Photosynthesis and growth. Fotosintez i produktsionnyi protsess (Photosynthesis and Productivity). Nichiporovich A.A., Ed., Moscow: Nauka, 109 p. (in Russian).

Nichiporovich, A.A. (1956) Photosynthesis and theory of obtaining high yields. Timiryazevskoe Chtenie, 15: 1-93 (in Russian).

Nichiporovich, A.A. (1982) Physiology of Photosynthesis and Plant Productivity. In: Fiziologiya fotosinteza (Physiology of Photosynthesis), Moscow: Nauka, 7 p.

Pleshkov, B.P. (1976) Practical works on plant biochemistry. Moscow: Kolos, 254 p. (in Russian). Плешков Б.П. Практикум по биохимии растений// М.: Колос, 1976. - 254 с

Pronina, N.A., Semenenko, V.E. (1984) Localization of membrane-bound and soluble forms of carbonic anhydrase in Chlorella cells. *Fiziol. Rast.,* 31: 241-251 (in Russian).

Raines, C.A. (2011) Increasing photosynthetic carbon assimilation in C3 plants to improve crop yield: current and future strategies. *Plant Physiology,* 155: 36-42.

Rasulov, B.Kh., Asrorov, K.A. (1982) Dependence of photosynthesis on specific leaf weight in different species of cotton plant. In: Fiziologiya fotosinteza (Physiology of Photosynthesis), Moscow: Nauka, 270 p.

Rasulov, B.Kh., Laisk, A.Kh., Asrarov, K.A. (1983) Photosynthesis and photorespiration in ontogeny of some cotton cultivars. *Fiziol. Rast.,* 30: 637-645 (in Russian).

Richli, EE., Chazanfar, S.A.S., Gibbons, B.H., Edsall, J.T. (1964) Carbonic anhydrase from human erythrocytes. Preparation and properties of two enzymes. *J. Biol. Chem.,* 239: 1065-1078.

Romanova, A.K. (1980) Biochemical methods of autotrophic study in microorganisms. Nauka, Moscow: 160 p. (in Russian).

Sakharova, O.V., Bykov, O.D., Verzhuk, V.T. (1984) Energetic and assimilation ability of spring winter leaves in ontogeny of plants with different length of vegetation period. *Trudy prikl. Bot. Genet. Selekts.,* 84: 83-89 (in Russian).

Šesták, Z., Jarvis, P.G., Catsky, J. (1971) Criteria for the selection of suitable methods. In: Plant photosynthetic production. Manual of methods (Šesták Z., Jarvis P.G., Catsky J., eds.). W. Junk Publishing Co., Hague: 1-48.

Thomas, S.M., Hall, N.P., Merrett, M.J. (1978) Ribulose-l,5-bisphosphate carboxylase/oxygenase activity and photorespiration during the ageing of flag leaves of wheat. *J. Exp. Bot.,* 29: 1161-1168.

Volodarskii, N.I., Bystrykh, E.E., Nikolaeva, E.K. (1980) Energetic effectiveness of photosynthesis in highly productive winter wheat cultivars. Nauch. Dokl. Vssh. Shkoly (biol. nauk.), 9: 84-90 (in Russian).

Voznesensky, V.L., Zalenskii, O.V., Semikhatova, O.A. (1965) Methods of plant photosynthesis and respiration researches. Nauka, Leningrad: 305 p. (in Russian).

Voznesensky, V.L. (1977) Photosynthesis of desert plants. Nauka, Leningrad, USSR: 256 p. (in Russian).

Wibur, K.M., Anderson, N.G. (1948) Electrometric and calorimetric determination of carbonic anhydrase. *J. Biol. Chem.,* 176 (1): 147.

In: Crop Yields
Editors: Lijuan Huang and Qiao Zhao
ISBN: 978-1-62618-175-5

Chapter 2

OPTIONS FOR OPTIMISED USE OF WATER RESOURCES FOR SECURING FUTURE AGRICULTURAL PRODUCTION

Christian Richardt Jensen*, Sven-Erik Jacobsen, Søren T. Jørgensen and Fulai Liu

Department of Plant and Environmental Sciences, Faculty of Sciences, University of Copenhagen, Taastrup, Denmark

HIGHLIGHTS

- Less furrow irrigation and more pressurized irrigation has a large potential for water saving.
- New insights and modelling tools for irrigation scheduling and fertigation save water and increase yield.
- Increased use of saline and waste water is potentially able to be handled in the future.
- New cropping systems with resistant crops should include optimised regional virtual water export/import strategies.
- Reduction of waste of food, feed and bio-fuel has a considerable potential for agricultural water saving.

* E-mail crj@life.ku.dk (C.R. Jensen); Phone: 0045 35 33 33 92; FAX: 0045 35 33 34 78.

ABSTRACT

Agriculture is the main user of fresh water resources in the World. However, irrigation water has become less available in many areas due to an increased competition for water among industries, households and the environment and due to global climate change. Thus, there is a growing interest of using irrigation water resources in a more efficient manner. Here a viewpoint is made by looking on the efforts and knowledge already existing about optimising the use of irrigation water resources. Main tools for optimising the use of water resources in agriculture can be obtained by: selection of water efficient irrigation methods like drip and micro irrigation, use of low quality waters such as treated wastewater, use of salt tolerant crops, optimising irrigation scheduling by using plant and soil sensors, remote sensing, irrigation models, using the 'virtual water' principles so that water rich regions secure food supply to dry regions. In dry regions small areas with high value crops should be irrigated securing net income and livelihood. In general water can be saved by reducing the overall waste of food throughout the food chain from post-harvest to the end-consumer. Recent findings on these options are discussed and evaluated.

Keywords: Pressurized deficit irrigation; irrigation sheduling; low quality water; cropping systems and 'virtual' water; waste of food, feed and bio-fuel

1. INTRODUCTION

Worldwide more than 40% of the food production comes from irrigated land and agriculture is the biggest consumer of water accounting for some 70-90 % of all freshwater withdrawals (EIIE, 2000; OECD/FAO, 2011). Three levels for decisions can improve awareness about water use (Hoekstra, 2003). The first level is the user level where increasing user's awareness, applying water pricing and water saving technology can lead to improvement in local water use efficiency. The second level is the river basin level. In this level, evaluation of the value of water for different alternative uses has to be done in order to reallocate water in an economically efficient way and hence improve water allocation efficiency. The third level is the global level. The application of virtual water trade between water scarce and water abundant regions can achieve improvements in the global water use efficiency (Horleman and Neubert, 2006). Arid countries can import food instead of extending irrigation;

i.e. import 'virtual water' (Chapagain et al., 2005), as a part of Integrated Water Resource Management for these countries (El-Sadek, 2011). Therefore water limitation should result in rethinking regional choices of cropping patterns and irrigation systems with consideration of the 'virtual water' concept. Tightly connected to the virtual water concept is the water footprint concept, which can be broken down to blue (surface and ground water), green (evaporated from soil stored water) and grey (polluted water) water footprint. The water footprint may be modified and reduced by reducing waste of food and feed (Gustavsson et al., 2011) and by rethinking the food composition so that food contain less products of animal origin with high footprint values (Hoekstra, 2012). Recently, a Global Water Footprint Standard has been launched (Hoekstra et al., 2011).

Improved management is most likely the best option in most agricultural systems for increasing the efficiency of water use (Steduto et al., 2007). As a framework for water saving, Hsiao et al. (2007) suggested working on a chain of efficiencies, i.e.: Conveyance efficiency (E_{conv}) and farm efficiency (E_{farm}), until the arrival of the water at the edge of the field. For the different ways of water distribution (furrow, sprinkler, surface drip and subsurface drip, with the latter eliminating most of the soil evaporation) field application efficiencies (E_{appl}) of 60, 75 and 90% for surface irrigation (border, furrow, basin), sprinkler irrigation and drip irrigation, respectively, have been obtained under European conditions (Dworak et al., 2007). Appropriately managed, subsurface drip irrigation (SDI) has recorded irrigation efficiency (IE) of as much as 95–100% (Schneider and Howell, 2001). SDI together with fertigation is applicable for a large number of crops (Camp, 1998; Hagin et al., 2003).

Irrigation often induces salinity due to dissolved salts in the irrigation water. When plants use water, the salts are left behind in the soil and eventually begin to accumulate. To avoid negative effects of salt accumulation the salts must be leached out of the root zone by applying excessive water defined as the leaching fraction. Poor drainage and use of saline brackish water for irrigating crops may exacerbate the salinisation. However, saline soils also represent a potential reserve for food production.

A key study by Fisher et al. (2002) found that more than three quarters of the global land is not suitable for rain-fed crop cultivation being too cold (13%), too dry (27%), too steep (12%), and some areas has poor soil conditions (65%). Note that percentages do not sum up to 100, because several constraints coincide in some regions. These authors conclude that expansion of agricultural land is not possible, so intensification of agriculture will be the most likely means to meet the food needs of 9.3 billion people in 2050. The study asserts that enough food can be produced on currently cultivated land if sustainable management and adequate inputs are applied. Projected climate change will cause mixed and geographically varying impacts on crop production.

The aim of the present review is to point out some options for saving irrigation water by: 1) adopting more water saving irrigation systems; 2) increasing efficient control of irrigation water in the field; 3) managing salinity and use of waste water for irrigation; 4) optimizing cropping patterns and including the principles of 'virtual water' and 'water footprint' concepts in regional cropping choices and 6) reducing the amount of waste of food, feed and bio-fuel.

2. Water Saving Potential by Changing Irrigation System

By using information collected in the framework of FAO's AQUASTAT program, Siebert et al. (2005) has developed maps of global irrigation areas. Siebert and Döll (2007) estimated that 274 million hectares (Mha) are areas equipped for irrigation (AEI), and 315 Mha irrigated area are harvested (IAH) around year 2000. The main irrigated crops at the global scale are rice (IAH of 103 Mha), wheat (IAH of 67 Mha), maize (IAH of 29 Mha) and cotton (IAH of 16 Mha). More than 50% of the global area with irrigation was located in India, China and in the United States. The baseline scenario for the International Water Management Institute (IWMI) assumes an increase in AEI of 22% from 1995 to 2025 (Seckler et al., 2000). For developing countries FAO estimates AEI to increase by 20% from 202 Mha (1997) to 242 Mha in 2030 (Bruinsma, 2003).

Many countries, such as Turkey, India, and China revised their irrigation legislation, subsidizing their farmers for water and energy saving irrigation equipment. At present, the global coverage of sprinkler and micro irrigated areas are 33 and 6 Mha, respectively (Sener, 2010). Hereby the water saving potential is huge if part of the surface irrigated land (estimated surface irrigated land 315-39 = 276 Mha) is reduced and turned into sprinkler and micro irrigated land.

3. IRRIGATION WITH SALINE WATER AND DRAINAGE OF IRRIGATED LAND

About 20-30 Mha, 10% of the irrigated land, are subject to salinization (Tanji and Kielen, 2002). Therefore drainage development is needed in places where the groundwater is saline. It is estimated that 10-20% of the irrigated land is already equipped with drainage systems; 20-40% of the irrigated area is not affected, while 40-60% is in need of drainage in arid and semi-arid areas of the world in South Asia, Southeast Asia, Central Asia, North Africa, the Near East, Australia and USA (Smedema et al., 2000). In order to avoid the negative effects of salts in drainage water there are four measures which can be used for drainage water management, i.e.: water conservation, drainage water reuse, drainage water treatment, and drainage water disposal. The former of these also has the potential for reducing use of freshwater resource for irrigation. The first line of action in on-farm water conservation is source reduction or reducing deep percolation. Where the goal is not achievable by source reduction, it may be necessary to implement other conservation measures in combination with reuse measures. As different crops require different amounts of water for optimal crop growth and as rooting patterns differ among crops, source reduction and water table management could be combined with a change in cropping pattern to optimize the desired effect of the measures. Changes in cropping pattern should only be considered where compatible with the broader development objectives for the area under consideration (Tanji and Kielen, 2002).

Salt concentration can be expressed as total dissolved solids (TDS) in mg of salt per liter of water (mg L^{-1}) or grams of salt per cubic meter of water (g m^{-3}) (i.e. ppm). Salt concentration can also be measured by the electrical conductivity of irrigation water (ECi). Electrical conductivity is normally expressed in millimhos per centimeter (i.e. ms cm^{-1}) now replaced by deciSiemens per metre (dS m^{-1}) or microSiemens per centimetre (μS cm^{-1}; i.e. 1dS/m = 1000μS/cm). The relationship between salt concentration (C; mg L^{-1}) and electrical conductivity (EC; dS m^{-1}) is approximately C = 640 EC. Another way of estimating salt concentration is by measuring the electrical conductivity of water extracted from a saturated soil sample (ECe). The approximate relationship between the electrical conductivity of irrigation water (ECi) and soil salinity is ECe = 1.5 ECi, if about 15 percent of the applied water is draining from the crop root zone. Under arid conditions both the quality of the irrigation water and soil salt content is important. The suitability of water reuse for irrigation with high salt concentration depends on salt tolerance of the crop, soil characteristics, climatic conditions, and soil and water management practices. The quality of the irrigation water plays an essential role in arid areas affected by high evaporation rates and causes high concentrations of salt accumulating in the soil. Irrigation water must have a low to medium salinity level (i.e. EC_i of 0.6 to 1.7 dS m^{-1} with a salt content less than 1000 mg L^{-1}) including the relative concentration of sodium to calcium and magnesium (Sodium Adsorption Ratio, SAR)<6. When $EC_i > 3.0$ dS m^{-1} the hazard of using the irrigation water is severe. Water with high saline (EC_i >1.5) and sodium (SAR > 6) concentrations should only be used for supplemental irrigation and therefore a good management and crop salt tolerance is important. If water with a very high salinity is used (e.g. at extreme water shortage circumstances), the soil must be permeable, drainage must be adequate meaning that water must be applied in excess to provide considerable leaching and salt-tolerance crops should be selected. Also, due to the additive effect of osmotic and matric potential during soil drying on soil water availability crops should be re-irrigated at higher soil water content in salt affected soils (Razzaghi et al., 2011). If a farmer annually applies 10,000 tons of irrigation water to 1 ha of crops, which is typical, between 2 and 5 tons of salt will be added to that land every year. Unless these salts are flushed out, enormous quantities can build up over the course of years. The salt concentration (C_L) of the percolation or leaching water is a function of the salt concentration of the soil solution in the root zone. The irrigation requirement (I_g) for salinity control can be calculated from actual evapo-transpiration from the root zone (E_{ra}), salt concentration of irrigation water (C_{ir}) and of the water of the root zone (C_r) and precipitation (P_p) as (Oosterbann, 2012):

$$I_g = \frac{(E_{ra} - P_p)\, f(C_r)}{f(C_r) - C_{ir}} \quad (1)$$

More mathematical, numerical models such as SaltMod (Oosterbaan, 2012) and SALTMED (Ragab et al., 2005) have been developed for the simulation and prediction of the salinity of soil moisture, ground and drainage water, the depth of the water table, drain discharge and leaching of salts in irrigated agricultural lands under different geohydrologic conditions, varying water management options, including the use of groundwater for irrigation by pumping from wells , and several crop rotation schedules.

Controlling salinity by drainage needs a leaching fraction of the total irrigation water to be 10-20% and the salt concentration of the drainage water is normally 5 to 10 times higher than that of the irrigation water resulting in the fact that salt content in the soil is not accumulating (Hoorn and Alphen, 2006). In order to save irrigation water in regions with distinct dry and wet seasons, the drainage system may be operated in the wet season only, reducing environmental hazards from salt in the leaching water. When drip irrigation is applied in orchards multiple lines of low-flow drip tape may be used where there is salt accumulation along the tree row, as opposed to putting water on the entire area of the field. This technique offers the potential for significant water savings for reclamation leaching (Burt and Isbell, 2005).

4. Pressurized Irrigation

The water saving pressurised irrigation systems (drip and sprinkler) have been compared with furrow irrigation systems (Burt et al., 1999; Raine and Foley, 2002). From such comparisons distinct differences appears between drip irrigation, sprinkler irrigation and pivot irrigation.

5.1. Drip Irrigation

With drip irrigation water is delivered close to the root systems and is in this way the most water-efficient method of irrigation minimizing evaporation and runoff. The high-tech solutions are controlled from computers with calibrated emitters supplied from computerized set of valves. Further reduction of water loss can be obtained from subsurface drip irrigation (SDI). Due to its high effectiveness in respect to avoid evaporative losses to the atmosphere and its environmental friendliness causing less soil erosion, SDI has received increased research effort in the last 30 years (e.g. reviewed by: Ayers et al., 1999; Camp, 1998; Camp et al., 2000). SDI has until now been limited to high value horticultural crops due to high cost of installation. However, worldwide irrigation manufactures in cooperation with irrigators are developing more simple and cheap solutions so that SDI is becoming more applicable for field row crops. A review on corn in Kansas (Lamm and Trooinen, 2003) indicates that SDI can save as much as 35-55% of the irrigation water compared with common irrigation practices in the region. Further nitrogen fertigation was a very effective management tool in combination with SDI resulting in high nitrogen and water use efficiencies. For cost effectiveness the systems should last 10-20 years.

5.2. Sprinkler Irrigation

Sprinklers, sprays, or guns can be mounted overhead permanently installed as solid-set irrigation systems or they can be mounted on automatically moving wheeled systems known as traveling sprinklers. The latter system may irrigate unattended.

5.2.1. Pivot Irrigation

There are three main types of pivots. The *Standard Central Pivot* is a movable irrigation system in which a lateral rotates around a fix point. At this fix point the system is fed with water, which is normally pumped from a reservoir outside the system. The *Towable Central Pivots* are normally smaller than standard central pivots. The *Lateral Move Pivots* are moved frontward while irrigating and are commonly used in rectangular surfaces. They can include a turning system in order to irrigate "L" shaped plots. They are more complex to operate as they need to be water-fed from external sources of water such as a lateral canal or long hoses. Hydrants should be placed every 200 metres and the hose should be reinstalled permanently. When irrigating with a good pivot with the right sprinklers irrigation uniformity up to 90-95% can be achieved.

By 2003 a huge increase in irrigation in USA had occurred to an irrigated area of 21.6 Mha, and in which 51% were irrigated by pivots, 43% by gravity and 6% by other technologies (micro-irrigation). In 2002 there were 170,000 pivot machines, irrigating at least 8 Mha in USA determined by land satellites system. The system also estimated that there were 88,000 pivots outside USA. Altogether, this industry expands between 7 and 11% each year. Between the late 1960s and early 80s there were 80 companies manufacturing pivots worldwide, predominantly in the USA (New Ag International, 2008). This major shift was mainly driven by the huge savings in labour force that pivots permit, but also in savings of water and energy. Marginal soils such as sandy soils or soils with slopes cannot be irrigated by gravity, but they might be irrigated by pivots. The biggest problem nowadays for the expansion of this technology is that the farm sizes are either too small or not suited for pivots in many countries. For a 50 ha cotton field size grown under Australian conditions Raine and Foley (2002) estimated that the costs of surface, pivot and drip irrigation is 1.159, 2.300, and 4.000 \$ ha^{-1}, respectively, indicating that the expenses per ha for pivot irrigation is about the half of those of drip irrigation. Crops are often planted in a circle to conform to the center pivot. Many center pivot systems now have drops hanging from a u-shaped pipe attached at the top of the pipe with sprinkler heads that are positioned about 1m (at most) above the crop, thus limiting evaporative losses. Originally, most center pivots were water powered. These were replaced by hydraulic systems (*T-L Irrigation*) and electric motor driven systems (Reinke, Valley, Zimmatic). Many modern pivots now feature GPS devices. The type of Pivot system known as LEPA (Low Energy Precision Application) has been developed in the USA to conserve both water and energy (Lyle and Bordovsky, 1981). This technology has been used primarily for row crop irrigation with application efficiencies typically exceeding 95% (Fipps and New, 1990). The mechanical component of a LEPA irrigation system is a moving truss system with water conveyance tubes extending from the system mainline to near the soil surface, where correctly sized orifices control deposition of water to individual soil furrows.

5. Scheduling Irrigation

By irrigation scheduling frequency, quantity and timing of irrigation is undertaken in order to optimize crop growth and production. Worldwide the irrigation is often based on climatic data roughly by calculating accumulated deficit (D) of daily water balance data from evapotranspiration (ET) and rainfall (R) [i.e. $D=\sum(R-ET)$]. During a drying cycle a certain D level will initiate an irrigation event (Hillel et al., 1998). Availability of weather data allows the use of advanced methods for calculation of ET predictions as reviewed by Allan et al. (1999). The amount of water applied is determined by using a threshold to determine irrigation need and a strategy to prescribe how much water to apply in any situation. Many models is based on knowing the current crop stage, soil water content and ET (Allen et al., 1999) and are therefore able to predict when to irrigate a crop (e.g. Steduto et al., 2007). Also crop water status (Jones, 2004) and soil water status (Steduto et al., 2009) may be used for irrigation scheduling. Irrigation scheduling is now common practise used by farmers and/or by advisers.

5.1. Use of Low Quality Waters and Fertigation

Low quality waters as waste water often contain nutrients. As irrigation and fertilization are the most important tools for farmers to control crop growth the use of low quality waters for irrigation has to be seen in the context of fertigation. The control of fertigation is superior under microirrigation (drip and microsprinkler). The principles of fertigation have been thoroughly reviewed by Bar-Yosef (1999). Fewer systems include nutrients and fertigation than for irrigation (e.g. Anastasiou et al., 2009; Heidman et al., 2008). So far no models for calculation of crop water requirement incorporate the hazards related to the use of wastewater. However, Styczen et al. (2010) included fertigation in a management model for decision support when applying low quality waters in irrigation. The cornerstone in the model is the deterministic "Plant–Soil–Atmosphere" model DAISY (Abrahamsen and Hansen, 2000), which simulates crop growth, water and nitrogen dynamics and if required heavy metals and pathogen fate in the soil. The irrigation and fertigation module calculates irrigation and fertigation requirements based on DAISY's water and nitrogen demands. A Water Source Administration module keeps track of water sources available and their water quality, as well as water treatments, storage, and criteria for selection between different sources. The management model system can be used for analysis prior to investments or when preparing a strategy for the season.

5.2. Sensors for Irrigation Control

Soil moisture sensors buried e.g. 20 cm deep within the crop root zone has been used to maintain soil moisture at a set level. The irrigation is starting if the soil volumetric water content or soil tension is below a set level. Such systems are now trivial and well established in advanced irrigation setups. The use of plant sensors for irrigation scheduling in orchard and field crops is under development. The use of plant based methods such as (dendrometry, fruit gauges, and other tissue water content sensors, measurements of growth, sap flow, and stomatal conductance) are thoroughly outlined for irrigation scheduling by Jones (2004). . Lately, Al-Yahyai (2012) found that in apple trees stem water potential was highly correlated with soil water potential showing potential possibilities of irrigation scheduling from real time measurement of stem water potential. A system with wirelessly transmitted information of several sensors appropriately distributed over various sectors of a round field irrigated by a center-pivot irrigation system could tell the irrigation lever exactly when and which field sector needs to be irrigated By remote sensing it is possible to obtain information about crop water status. Santos et al. (2010) made a remote sensing-based assessments of actual ET using METRIC integrated with a water balance model providing accurate estimates of irrigation performance. The proposed methodology allows the estimation of the volume of applied water at the field scale. Comparisons of values obtained from actual applied water records against those calculated with the new methodology resulted in good agreement. METRIC estimates ET as a residual of the energy balance at the surface and is calculated as:

$$LE = Rn - G - H \tag{2}$$

where LE is the latent energy consumed by ET, Rn is net radiation, G is sensible heat flux conducted into the ground, and H is sensible heat flux to the air. Details of the METRIC model are given by Allen et al. (2007), and the model was tested by Santos et al. (2010). Recently, Ha et al. (2012) reviewed downscaling methods to improve spatial resolution of land surface characteristics such as surface temperature and daily ET.

6. Deficit Irrigation

Deficit Irrigation (DI) is a method of irrigation where the amount of water used is kept below the potential ET and the minor stress that develops has minimal effects on the yield. In this mode of irrigation the entire root-zone is irrigated. Knowing when to apply water is a must for the successful implementation of DI because the sensitivity of the crop to water stress is different at different growing stages. There is, for example, a high sensitivity during flowering with a high risk of a yield decrease via flower or seed abortion (Kirda, 2002; Liu et al., 2004). Knowing the crop and soil water status in a certain growing environment is important to ensure the correct irrigation practice so water deficits can be imposed at times when it has minimum affect on yield, a practice also termed regulated deficit irrigation (RDI). RDI aims to use drought stress to control vegetative and reproductive growth, and to avoid any luxury water consumption. This requires careful control of the timing and level of water deficit. The response of plants to DI or RDI can be reflected in one factor or in a combination of stimulation of root growth, maintenance of high leaf water potential, osmotic regulation of leaf-turgor pressure, and adaptation to existing soil water status by e.g. reduced leaf elongation and stomatal closure. DI is more common in fruit trees and vines because economic returns in these crops are generally higher than in field crops and less directly related to biomass production (Hsiao et al., 2007) than to the quality of fruits and yield (Fereres et al., 2003). Nonetheless, in processing tomatoes DI (or RDI) is widely applied in the Emilia Romagna region, Italy, by irrigation extension service to save water on the basis of local research and using a decision support system (DSS) (Battilani et al., 2000; 2003). In drought sensitive species like potatoes and tomatoes extensive field studies showed that 20-25% of the total irrigation water could be saved by DI (Jensen et al., 2010) without yield depression as compared with fully irrigated plants. The later part of the growing season was the most robust period for saving irrigation water.

The root chemical signalling mediated by abscisic acid (ABA) for 'isohydric' plants most probably act as a feed-forward mechanism controlling stomatal aperture (Chaves and Oliveira, 2004). However, recent investigations in field grapevines suggest if severe soil drying saving 50% water of control plants is applied during deficit irrigation feedback hydraulic signals may play a dominant role (Rodrigues et al., 2008). The investigations of drought induced adaptation processes has a potential for understanding the mechanisms of deficit irrigation (Shao et al., 2008) and hereby its potential for water saving.

7. Water Footprint and Virtual Water and Land Savings

Water productivities, water management and virtual water flows are important to consider in order overcoming scarcity of water for a growing population in the future. Water footprint (WFP) and virtual water have lately been calculated for crops, goods, services, as well as on generic national levels (e.g. Hoekstra and Chapagain, 2007). The WFP of a product is defined as the total volume of fresh water used to produce the product, summed over the various steps in the production chain (e.g. Hoekstra et al., 2009).

About 3000 to 6000 km^3 yr^{-1} of "green" water (precipitation stored in the soil and evapotranspired on cropland) are consumed in addition to sustain rainfed agriculture and parts of irrigated agriculture. The global consumption of "blue" water (taken from rivers, reservoirs, lakes and aquifers and used for irrigation) presently amounts to 927–1660 km^3 yr^{-1} according to recent estimates (Rost et al., 2008; Hoff et al., 2010). Fader et al. (2011) for 11 of the world's major crop species modelled the amount of blue and green water (irrigation and precipitation water) and quantified WFP distinguishing internal WFP (local water) and external WFP (virtual water imported from other countries) and their blue and green components, respectively. Green water flows and footprints show that green water globally dominates both the internal and external WFP (84% of the global WFP and 94% of the external WFP rely on green water). Nevertheless, current trade of the products considered here saves significant water volumes (263 km^3) and land areas (41 Mha), equivalent to 5% of the sowing area of the considered crops and 3.5% of the annual precipitation on this area. Considering the land needed in order to produce imported goods on their own territory, Fader et al. (2010) estimate that China and Mexico would need 9 Mha each, North Korea and The Netherlands 7 Mha each, and Japan 16 Mha. Relating these needs to the current cropland extent demonstrates that many countries – 40 in total, especially in North Africa and Latin America – would have to more than double the current cropland to produce their imports on their own territory.

From the virtual water standpoint, soybean consumes 3.1 m^3 water kg^{-1} beans while cereals consume around 1.2–1.4 m^3 water kg^{-1} grain and maize 0.7 m^3 water kg^{-1} grain. Tomato and potatoes consume 0.1 m^3 water kg^{-1} fruit (Table 1). Aldaya et al. (2010) found that blue water economic productivity ranges between 0.1–0.2 euros m^{-3} for low-cost cereals and 2–3 euros m^{-3} for vegetables. Vegetables seem to be the most profitable crops, followed by grapes and olives.

Table 1. Average crop water requirement (modified after Yang et al., 2007)

Species	m^3/kg	Species	m^3/kg
Wheat	1.2	Sugarcane	0.3
Rice	1.4	Soybean	3.1
Maize	0.7	Beans	1.7
Barley	1.2	Groundnut	2.6
Millet	1.2	Tomatoes	0.1
Sorghum	0.7	Onions	0.1
Potatoes	0.1	Orange	0.4
Sugar beet	0.2	Apples	0.4
		Banana	0.5

8. Food, Feed and Bio-Fuel Production and Waste

Worldwide there is a growing demand for grain for satisfying the need for food, feed, and fuel. By increasing the production on the existing farmland it avoids greenhouse gas production by bringing new land and ecosystems into production. In USA average corn yield has increased from 1.6 ton ha^{-1} in the first third of the 20th century to today's 9.5 ton ha^{-1}. This increase is due to new farming technologies such as hybrid corn, use of fertilizers and machinery (Edgerton, 1999). With an increase in prosperity people consumes more meat and dairy products every year. Global meat production is projected to more than double from 229 Mt in 2000 to 465 Mt in 2050 and a similar increase is projected for milk from 580 to 1043 Mt. (OECD/FAO, 2011). The increase in middle class consumer demand shifts from stable foods towards processed food products with more animal protein. Based on OECD/FAO predictions global consumption of wheat is expected to decline over the next 10 years, rice production is still to be increased for use in Asia. However grains for feed (poultry, pork), and sugar cane and oil crops for bio-fuel is expected to increase (Table 2).

Food and feed use still dominates cereal production.. In 2012 cereal consumption for food, feed, biofuel and other purposes are c. 1150, 800, 170, and 200 Mt, respectively, and are anticipated to increase by 25% within the next decade (OECD/FAO, 2011). The demand for grain is a result of the increasing world population (Friend, 2009). Due to climate change the areas useful for grain production is expected to change in the coming years especially expanding into the northern part of the world with sufficient rain fall (see Plates K and L given by Fischer et al., 2002).

Large losses and waste occur from food and feed products from the farmer's field until it reaches its final destination; in storage, transportation, processing, wholesale and retail (Table 3).In 2011 it was estimated to be 1.3 billion tons of food or about 1/3 of the global food production was wasted (Gustavsson et al., 2011). Furthermore, vast amounts of food are then wasted in households and restaurants. Every loss or waste of food is equivalent to loss and misuse of water and waste of land (Table 4).

Animal-based food requires a larger amount of water per kilo ready-to-eat product than do fruits and vegetables increasing the amount of water wasted. A 50 percent reduction of food losses and waste at the global level would save 1 350 km^3 of water a year (OECD/FAO, 2011).

Table 2. Predicted growth in *per capita* consumption of food products (2008-10 to 20). (After OECD/FAO, 2011)

Product	Percent	Product	Percent
Wheat	-0.3	Sheep	8.7
Rice	6.0	Fish	4.8
Coarse grains	6.0	Milk	9.4
Beef	1.0	Veg. oils	10.6
Pork	6.3	Sugar	14.4
Poultry	14.4		

Table 3. Waste percentages for each commodity group in each step of the food supply chain for Europe incl. Russia (Partly after Gustavsson et al., 2011)

	Agricultural production	Postharvest handling and storage	Processing and packaging	Distribution: Supermarket Retail	Consumption	Total
Cereals	2	4	0.5	2	25	33.5
Roots and Tubers	20	9	15	7	17	68
Oilseeds and Pulses	10	1	5	1	4	21
Fruit and vegetables	20	5	2	10	19	56
Meat	3.1	0.7	5	4	11	23.8
Fish and Seafood	9.4	0.5	6	9	11	35.9
Milk	3.5	0.5	1.2	0.5	7	12.7

Table 4. Estimated water lost (km^3) due to waste percentages for each commodity group in each step of the food supply chain (FSC) for Europe incl. Russia. (Partly after Gustavsson et al., 2011* and Hoestra, 2012)**

	*Commodity (million tonnes)	**Water footprint, m^3/tonnes	Water, km^3
Cereals	400	1640	656
Roots and Tubers	180	390	70
Oilseeds and Pulses	175	3200	560
Fruit and vegetables	201	900	181
Meat	45	8300	374
Fish and Seafood	Not estimated	Not estimated	
Milk	232	3000	696
Total			2537

DISCUSSION AND CONCLUSION

The continuous growing population in the world, with an increasing part being in the middle class, results in an increasing need for food, feed and bio-fuel production by about 25% within the next decade. However expansion of the cultivated agricultural land area beyond existing level seem not possible. Therefore, intensification and optimizing of the cultivation of the existing agricultural land is necessary. As argued by Fisher et al. (2002) establishment of more irrigated cultivated land is limited by lack of renewable ground water reserves and by soil salinization. This is in accord with estimates that in developed countries the number of dams for irrigation water remains stable, however in developing counties a limited number new reservoirs will be built in the future, but with a small number compared to the existing 45,000 large dams (Scudder, 2005). Therefore refinement of the present agricultural systems is needed to meet the higher grain production required in the future. Several options for optimizing agricultural water resources under arid conditions are obvious to take into account. Firstly, an accelerated shift from surface irrigation (276 Mha) towards drip and sprinkler irrigation (39 Mha) is needed. This may save 20-30% of irrigation water which can be used for sprinkler irrigation of grain cultivation as supplemental irrigation (Jacobsen et al., 2012). Center pivot irrigation systems are promising for grain production. However, they need big fields to be implemented. Therefore they are most useful for larger landowners and farmers who are able to work in cooperatives with large field size. Further saving of water can be obtained if irrigation scheduling and deficit irrigation is implemented.

Globally mean precipitation will increase due to climate change. Current climate models tend to project increasing precipitation at high latitudes and less precipitation in some semi-arid and sub-humid regions (e.g. North Africa and the northern Sahara). Other parts of the world have already increasingly suffered from drought which may be worsened by climate change (Kundzewicz et al., 2007). Due to climate change the potential area for rainfed grain production is expected to move more North and South . Thus Canada, North Europe, Russia, South Asia, Central Africa and South America are predicted to gain precipitation and increase the potential for grain production (Kundzewicz et al., 2007). Scheduled supplemental sprinkler irrigation will under dry spells be able to increase yield by 30-40% (Fisher et al., 2002) maintaining yield close to its potential under fully irrigated conditions.

The effect of climate change on future grain production may thus establish large rainfed grain producing areas away from densely populated regions like

Central and South Europe, Middle East, West Africa, USA and North China. Due to the distance from the region where the grain is produced to where it is consumed as food, feed or bio-fuel makes the concept of virtual water even more important. However, new grain growing areas may also produce bio-fuel and feed consumed in the new regions. This is in accord with policy trends that Integrated Water Resources Management will be increasing around the world (World Water Council, 2006; Kundzwicz et al., 2007). However the largest future relative change in water demand per discharge has been forecasted to come from population growth and only in minor from climate change (Vörösmarty et al., 2010).

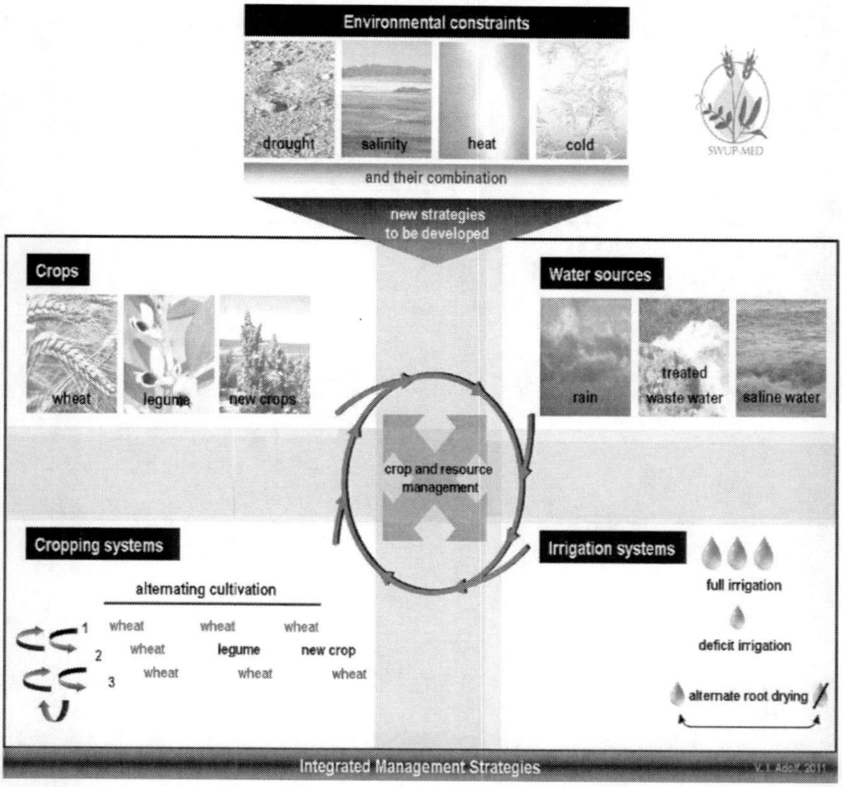

Figure 1. A schematic presentation of crop and resource management under the influence of multiple, environmental constraints (after Jacobsen et al., 2012).

Recently, Jacobsen et al. (2012) showed how different agronomic measures can be used to support plant production under semi arid conditions. Supplemental irrigation such as deficit irrigation has the potential to overcome periods of low rainfall or high temperatures. It is suggested that improvements in crop production may arise from several strategies such as early sowing enabled by minimum tillage, increased use of organic manure, and an efficient weed control. Further, crop rotations will play an important role in improving weed control, minimizing disease risk, and increasing nitrogen availability. Introduction of drought and salt tolerant crop species such as quinoa and amaranth may result in more resilient crop rotations and high value cash crop products. Genotypic increases of water use efficiency may arise from selection for early vigour, deep roots, increased transpiration efficiency, improved disease resistance, and high assimilate storage and remobilization. A range of crop and management strategies might be combined for a specific target environment in order to optimize crop productivity (Figure 1).

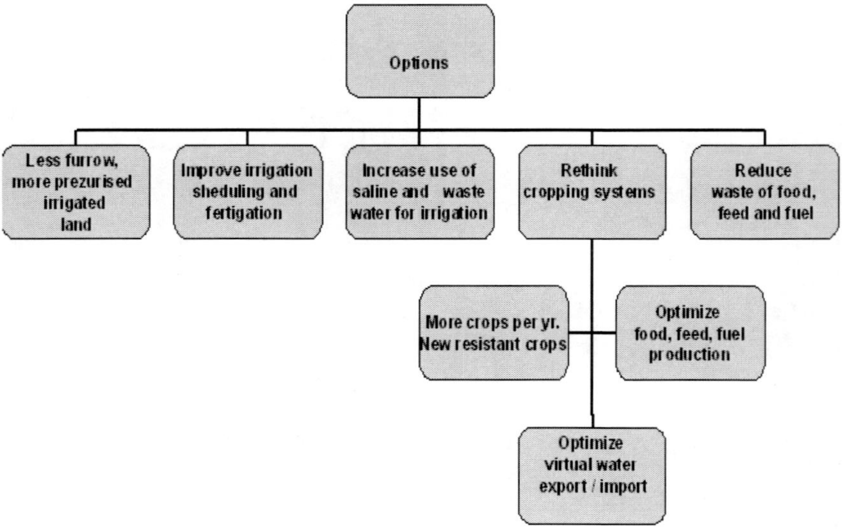

Figure 2. Options at farm, retail and consumer level for optimizing water resources for securing the future agricultural production.

From the above it can be concluded that the options for optimizing agricultural water resources are (Figure 2):

- More surface irrigated land is turned into sprinkler, pivot, or drip irrigated land.
- Irrigation scheduling, supplemental and deficit irrigation are taken into account and become common practice.

- Use of waste and saline water for irrigation is further developed.
- Imrovement of cropping systems and inclusion of saline soil soils for crop production using salt resistant crops and drainage systems for leaching of salts.
- Using the principles of water footprint and virtual water concepts for inter-regional planning of utilization of water resources.
- Reduction of waste of food, feed and bio-fuel.

REFERENCES

Abrahamsen, P. & Hansen, S.(2000). Daisy: an open soil–crop–atmosphere system model. Environ. *Model. Softw*. 15, 313–330.

Aldaya, M.M., Santos, M. & Llamas, R. (2010). Incorporating the Water Footprint and Virtual Water into Policy: Reflections from the Mancha Occidental Region, Spain. *Water Resour. Manage*. 24, 941–958.

Allen, R.G., Pereira, L.S., Raes, D. & Smith, M.(1999). *Crop evapotranspiration. FAO Irrigation and Drainage Paper* 56. Rome: FAO.

Allen, R.G., Tasumi, M. & Trezza, R. (2007). Satellite-based energy balance for mapping evapotranspiration with internalized calibration (METRIC)—model. *J. Irrig. Drain. Eng. ASCE* 133(4), 380–394.

Al-Yahyai, R.(2012) . Managing irrigation of fruit trees using plant water status. *Agric. Sci.* 3 (1), 35-43.

Anastasiou, A., Savvas, D., Pasgianos, G., Sigrimis, N., Stangellini, C., Kempkes, F.L.K. (2009). *Decision support for optimised irrigation scheduling*. In: Tuzel, Y., Oztekin, G.B. & Meric, M.K. (Eds.), *International Symposium on Strategies Towards Sustainability of Protected Cultivation in Mild Winter Climate*, pp. 253–258. ISHS Acta Horticulturae 807.

Ayars, J.E., Phene, C.J., Hutmacher, R.B., Davis, K.R., Shoneman, R.A., Vail, S.S. & Mead, R.M. (1999). Subsurface drip irrigation of row crops: a review of 15 years of research at the Water Management Research Laboratory. *Agric. Water Manage*. 42, 1–27.

Bar-Yosef, B. (1999). Advances in fertigation. *Adv. Agron*. 65, 2-77.

Battilani, A., Bussieres, P. & Dumas, Y. (2000). IRRIGERE: an improved version of an irrigation scheduling model for the processing tomato crop. *Acta Hort.* 537, 519-526.

Battilani, A., Dalla Costa, L. & Lovatti, L. (2003). Potato's efficient use of water and nitrogen in a sub-humid area. *Acta Hort*. 664, 63-70.

Bruinsma, J. (ed)(2003). *World agriculture towards 2015/2030 – an FAO perspective*. 444 p. FAO, Rome, Italy,.

Burt, C.M., Clemmens, A.J., Bliesner, R., Merriam, J.L. & Hardy, L. (1999). *Selection of irrigation methods for agriculture*. On-Farm Irrigation Committee, American Society of Civil Engineers, Reston, Virginia.

Burt, C.M. & Isbell, B. (2005). Leaching of accumulated soil salinity under drip irrigation. ITRC Paper No. P 05-001. *Transactions of the ASAE* , 48(6), 1-7.

Camp, C.R. (1998). Subsurface drip irrigation: A review. *Trans ASAE* 41 (5), 1353–1367.

Camp, C.R., Lamm, F.R., Evans,R.G. & Phene, C.J. (2000). *Subsurface drip irrigation: past, present and future*. In: *Proceedings of the fourth decennial irrigation symposium,* November 14–16, Phoenix, Arizona. pp 363–372. American Society of Agricultural Engineers, St Joseph, Mich., USA,.

Chapagain, A.K., Hoekstra, A.Y. & Savenije, H.H.G. (2005). *Saving water though global trade*. UNESCO-IHE, Value of Water Research Report Series No. 17, Delft, The Netherlands.

Chaves, M.M. & Oliveira, M.M. (2004). Mechanisms underlying plant resilience to water deficits:prospects for water-saving agriculture. *J. Exp. Bot*. 55, 2365–2384.

Dworak, T., Berglund, M., Laaser, C., Strosser, P., Roussard J., Grandmougin, B., Kossida, M., Kyriazopoulou, I., Berbel, J., Kolberg, S., Rodriguez-Diaz, J.A., Montesinos, P. (2007). *EU Water saving potential* (Part 1-Report), ENV.D.2/ETU/2007/001r.

Edgerton, M.D. (2009). Increasing crop productivity to meet global needs for feed, food, and fuel. *Plant Physiol*. 149, 7-13.

EIIE (Environmental Impacts of Irrigation in the European Union), (2000). *A report to the Environment Directorate of the EC*, March 2000.

El-Sadek, A.(2011). Virtual water: an effective mechanism for integrated water resources management. *Agric. Sci*. 2(3), 248-261.

Fader, M., Gerten, D., Thammer, M., Heinke, J., Lotze-Campen, H., Lucht, W. & Cramer, W. (2011). Internal and external green-blue agricultural water footprints of nations, and related water and land savings through trade. *Hydrol. Earth Syst. Sci.* 15, 1641–1660.

FAO (2000). FAO IRRIGATION AND DRAINAGE. PAPER 61. Hydrology Program Department of Land, Air and Water Resources. pp. 188. *FAO Land and Water Development Division*,.

Fereres, E., Goldhamer, D.A., Parsons, L.R. (2003). Irrigation water management of horticultural crops. *HortSci.* 38, 1036-1042.

Fipps, G.,& New, L.L. (1990). "*Six Years of LEPA in Texas - Less Water, Higher Yields. Visions of the Future*," *Proceedings of the Third National Irrigation Symposium.* 115-120. Phoenix, AZ, USA

Fisher, G., Velthuizen van, H., Shah, M. & Nachtergale, F. (2002). *Global Agro-ecological Assessment for Agriculture in the 21st Century: Methodology and Results.* pp 119. *IIASA*, FAO, Luxemburg, Austria. .

Friend, A.D. (2009). Darwin Review. Terrestrial plant production and climate change. *J. Exp. Bot.* 61 (5), 1293–1309.

Gustavsson, J., Cederberg, C., Sonesson, U., Otterdijk, R. van & Meybeck, A. (2011). *Global food losses and waste. Extent, causes and prevention.* Pp 12. FAO, Rome, Italy, Ha, W., Gowda, P.H.& Howell, T.A. (2012). A review of downscaling methods for remote sensing-based irrigation management: part I. *Irrig. Sci.* (DOI 10.1007/s00271-012-0331-7) published online.

Hagin, J., Sneh, M. & Lowengart-Aycicegi, A. (2003). *Fertigation. Fertilization through Irrigation.* . A.E Johnston (Ed.). *IPI Resesarch Topics* No. 23, 81 p.Internatinal Potash Institute, Basel.

Heidmann, T., Tofteng, C., Abrahamsen, P., Plauborg, F., Hansen, S., Battilani, A., Coutinho, J., Doleˇzal, F., Mazurczyk, W., Ruiz, J.D.R., Takaˇc, J., & Vacek, J. (2008). Calibration procedure for a potato crop growth model using information from across Europe. *Ecol. Model.* 211, 209–223.

Hillel, D. (1998). *Environmental Soil Physics.* Academic Press, New York.

Hoekstra, A.Y. (2003). Virtual water: An introduction. *Proceedings of the International Expert Meeting on Vitual Water Trade,* Delft, *Value of Water Research Report Series No. 12.*

Hoekstra, A.Y. (2012). The hidden water resource use behind meat and dairy. *Animal Frontiers* 2 (2). 3-8.

Hoekstra, A.Y. & Chapagain, A.K. (2007).Water footprints of nations: Water use by people as a function of their consumption pattern. *Water Resour. Manage.* 21, 35-48.

Hoekstra, A.Y., Chapagain, A.K., Aldaya, M.M. & Mekonnen, M.M. (2009). *Water Footprint Manual. State of the Art 2009; Water footprint Network*: Enschede, The Netherlands.

Hoekstra, A. Y., Chapagain, A.K., Aldaya, MM, & Mekonnen, M.M. (2011). *The Water Footprint Assessment Manual: Setting the Global Standard*. Earthscan, London, UK.

Hoff, H., Falkenmark, M., Gerten, D., Gordon, L., Karlberg, L. & Rockstrom, J. (2010). Greening the global water system. *J. Hydrol.* 384, 177–186.

Hoorn, J.W., van & Alphen, J. G. van (2006). *Salinity control*. In:, *Drainage Principles and Applications*, H.P. Ritzema (Ed.). Pp. 533-600, Publication 16, International Institute for Land Reclamation and Improvement (ILRI), Wageningen, The Netherlands. ISBN 90 70754 339.

Horlemann, L. & Neubert, S.(2006). *Virtual water trade: A realistic concept for resolving the water crisis?* German Development Institute, Bonn, Germany. – (Studies / Deutsches Institut für Entwicklungspolitik; 25). ISBN 978-3-88985-335-6.

Hsiao, T.C., Steduto, P. & Fereres, E. (2007). A systematic and quantitative approach to improve water use efficiency in agriculture. *Irrig. Sci.* 25, 209-231.

Jacobsen, S.-E., Jensen, C.R. & Liu, F. (2012). Improving crop production in the arid Mediterranean climate. *Field Crops Res.* 128, 34–47.

Jensen, C.R., Battilani, A., Plauborg, F., Psarras, G., Chartzoulakis, K., Janowiak, F., Stikic, R., Jovanovic, Z., Li, G., Qi, X., Liu, F., Jacobsen, S.E. & Andersen, M.N.(2010). Deficit irrigation based on drought tolerance and root signalling in potatoes and tomatoes. *Agric. Water Ma*nage. 98, 403–413.

Jones, H.G. (2004). Irrigation scheduling: advantages and pitfalls of plant-based methods. *J. Exp. Bot.* 55 (407), 2427–2436.

Kirda, C. (2002). Deficit irrigation scheduling based on plant growth stages showing water stress tolerance. In: *Deficit irrigation practices*. Pp. 3-10. FAO, Rome,.

Kundzewicz, Z.W., Mata, L.J., Arnell, N.W., Döll, P., Kabat, P., Jimenez, B., Miller, K.A., Oki, T., Sen, Z. & Shiklomanov, I.A. (2007). *Freshwater resources and their management. Climate Change 2007: Impacts, Adaptation and Vulnerability. Contribution of Working Group II to the Fourth Assessment Report of the Intergovernmental Panel on Climate Change*. ML Parry, OF Canziani, JP Palutikof, PJ van der Linden, CE Hanson (Eds.), pp 173-210. Cambridge University Press, Cambridge, UK, Lamm, F.R. & Trooien, T.P.(2003). Subsurface drip irrigation for corn production: a review of 10 years of research in Kansas. *Irrig. Sci.* 22, 195-200.

Liu, F., Jensen, C.R. & Andersen, M.N. (2004). Pod set related to photosynthetic rate and endogenous ABA concentration in soybeans subjected to different water regimes and exogenous ABA and BA at early reproductive stages. *Ann. Bot.* 94, 405-411.

Lyle, W.M. & Bordovsky, J.P. (1981). Low Energy Precision Application (LEPA) Irrigation System. *Transactions of the ASAE* 24 (5), 1241-1245.

New Ag International, 2008. Pivot Irrigation for increasingly sophisticated machines. *New Ag International*, 31-41.

OECD/FAO 2011. OECD-FAO *Agricultural Outlook* 2011-2020, Chapter 1 Overview, pp. 17-50. OECD Publishing and FAO.<http://dx.doi.org /10.1787/agr_outlook-2011-en.>.

Oosterbann, R.J. (2012). *SaltMod: agro-hydro-salinity model for drainage, leaching of salts, and land rehabilitation*. <http://www.waterlog.info/ saltmod.htm>.

Ragab, R., Malash, N., Gawad, A. G., Arslan, A. & Ghaibeh, A. (2005). A holistic generic integrated approach for irrigation, crop and field management: 2. The SALTMED model validation using field data of five growing seasons from Egypt and Syria. *International Journal of Agric. Water Manage*. 78 (1-2), 89-107.

Raine, S.R. & Foley, J.P. (2002). *Comparing application systems for cotton-What is the pros and cons*. <http://www.usq.edu.au/users/raine/index_files /Raine and Foley Nat Cotton Conf2002.pdf>.

Rodrigues, M.L, Santos, T.P., Rodrigues, A.P., de Souza, C.R., Lopes, C.M., Maroco, J.P., Pereira, J.S. & Chaves, M.M. (2008). Hydraulic and chemical signalling in the regulation of stomatal conductance and plant water use in field grapevines growing under deficit irrigation. *Funct. Plant Biol*. 35, 565–579.

Razzaghi, F., Ahmadi, S.H., Adolf, V.I., Jensen, C.R., Jacobsen, S.-E. & Andersen, M.N.(2011). Water relations and transpiration of quinoa (Chenopodium quinoa Willd.) under salinity and soil drying. *J. Agron. Crop Sci.* 197 (5), 348-360.

Rost, S., Gerten, D., Bondeau, A., Lucht, W., Rohwer, J. & Schaphoff, S. (2008). Agricultural green and blue water consumption and its influence on the global water system. *Water Resour. Res*. 44, W09405, 17 PP. doi:10.1029/2007WR006331.

Santos, C., Ignacio J., Lorite, Tasumi, M., Allen, R.G. &Fereres, E. (2010). Performance assessment of an irrigation scheme using indicators determined with remote sensing techniques. *Irrig. Sci.* 28, 461–477.

Schneider, A. D., Howell, T. A. & Evett, S. R. (2001).Comparison of SDI. LEPA and spray irrigation efficiency. *ASAE paper* No. 012019, pp 12.

Scudder, T. (2005). *The Future of Large Dams*. pp 408. Earthscan, London,.

Seckler, D.U., Amarasinghe, U., Mold, D., De Silva, R. & Barker, R. (2000). *World water demand and supply 1990 to 2025 senarios and issues*. IWMI Research Report 19, pp 40, IWMI, Colombo, Sri Lanka.

Sener, S. (2010). Global Climate Change and its Effect on Irrigation in Mediterranean Region - Water (Energy Saving Irrigation Systems and Some Relevant Research Results). In: *CLIMWATER 2010: Horticultural Use of Water in a Changing Climate*. 28th International Horticultural Congress. August 22-27 2010, Lisboa.

Shao, H.B., Chu, L.Y., Jaleel, A., Zhao, C.X. (2008). Water-deficit stress-induced anatomical changes in higher plants. *Comptes Rendus Biologies* 331, 215-225.

Siebert, S., Döll, P., Hoogeveen, J., Faures, J.-M., Frenken, K., Feick, S. (2005). Development and validation of the global map of irrigation areas. *Hydrol. Earth System Sci.* 9, 535-547.

Siebert, S. & Döll, P. (2007). *Irrigation water use – A global perspective*. In: JL Lozan, H Grabl, P Hupfer, L Menzel, C-D Schönwiese (Eds.): *Global Change: Enough Water for all?* Universität Hamburg/ GEO, 104-107.

Smedema, L.K., Abdel-Dayem, S. & Ochs, W.J. (2000). Drainage and agricultural development. *Irrig. and Drainage Sys.*, 14 (3): 223-235.

Steduto, P., Hsiao, T.C. & Fereres, E. (2007). On the conservative behavior of biomass water productivity. *Irrig. Sci.* 25, 189–207.

Steduto, P., Hsiao, T. C., Raes, D. & Fereres, E. (2009). AquaCrop-The FAO Crop Model to Simulate Yield Response to Water: I. Concepts and Underlying Principles. *Agron. J.* 101, 426-437.

Styczen, M., Poulsen, R.N., Falk, A.K. & Jorgensen, G.H. (2010). Management model for decision support when applying low quality water in irrigation. *Agric. Water Manage.* 98, 472–481.

Tanji, K.K. & Kielen, N.C. (2002). *Agricultural drainage water management in arid and semi-arid areas USDA statistics*. <http://wqic.nal.usda.gov/nal_display/index.php?info_center=7andtax_level=1andtax_subject=586>

Yang, H., Wang, L. & Zehnder, A.J.B. (2007).Water scarcity and food trade in the Southern and Eastern Mediterranean countries. *Food Policy* 32, 585–605.

Vörösmarty, C.J., McIntyre, P.B., Gessner, M.O., Dudgeon, D. Prusevich, A. Green, P., Glidden S., Bunn, S.E., Sullivan, C.A. Reidy Liermann, C. &

Davies, P.M. (2010). Global threats to human water security and river biodiversity. *Nature* 467, 555-561.

World Water Council, 2006. *Final report of the 4th World Water Forum*, p 262.

National Water Commision of Mexico, Mexico City.

In: Crop Yields
Editors: Lijuan Huang and Qiao Zhao
ISBN: 978-1-62618-175-5

Chapter 3

CLIMATE CHANGE AND WHEAT CROP YIELDS IN AUSTRALIA: A REVIEW

Qunying Luo[*]
University of Technology, Sydney, NSW

ABSTRACT

The potential impact of climate change on wheat production is a key concern to Australia due to its importance to the nation's economy and to global food security as wheat is a key player in the world grain market. This work extensively reviewed the relationship between wheat crop production including yield and future climatic change based on both experimental and modelling studies conducted in Australia over the last four decades. Experiment studies evolved from adopting glasshouse, gradient tunnel facility to free air CO_2 enrichment facility, evolved from studying enhanced CO_2 effect alone to the combined effects of increased atmospheric CO_2 concentration along with other environmental factors such as light, temperature, soil water and nutrient, and with various soil types, and evolved from wheat monoculture to wheat-pulse rotation system. Modelling approaches evolved from sensitivity analysis, to scenario analysis and to Bayesian analysis. The advantage and disadvantage of each research method were discussed where appropriate. Future research directions were identified to better address future climate change, food security and sustainability issues.

[*] Corresponding author Email address: luo.qunying122@gmail.com.

Keywords: Wheat production, yield, climate change, mechanistic modelling, experimental study

1. INTRODUCTION

Climate change, induced by enhanced greenhouse gases (GHGs, i.e. CO_2, N_2O, CH_4) emission due to anthropogenic activities such as fossil fuel burning, land use change and cement production, has received great attention from government and research communities since 1970s. As a result, substantial funding has been allocated to detect, attribute, project and mitigate climate change, to understand its impact mechanism on various sectors/industries which underpins the identification of effective adaptation strategies in dealing with projected negative impact and in taking the advantage of its positive impact.

Wheat is the major crop in Australia in terms of value ($5.2 billion), volume (25 Mt) and area (12.5 Mha) in 2005–06 (ABARE, 2008). Its importance to the national economy manifests itself in two ways: the gross value and the export value. The gross value of wheat production in 2005-06 represented 15% of the total value of farm production. More importantly, wheat is one of Australia's most valuable exports, with overseas sales in 2005-06 of over $3.2 billion, making it a significant player in the world wheat market (ABARE, 2008). The potential impact of climate change on wheat production is a key concern to Australia due to its importance to the nation's economy and to global food security as wheat is a key player in the world grain market. The potential impacts of future climate change including atmospheric CO_2 concentration (pCO_2) on Australian wheat production have been extensively studied in the last four decades by adopting both experimental studies and process-oriented modelling studies. Experiment studies evolved from adopting glasshouse, gradient tunnel facility to free air CO_2 enrichment (FACE) facility, evolved from studying enhanced CO_2 (eCO_2) effect alone to the combined effects of eCO_2 along with other environmental factors such as the levels of light, temperature, soil water and nutrient, and with various soil types, and evolved from wheat monoculture system to wheat-pulse rotation system. Modelling approaches evolved from sensitivity analysis (from single level of a single factor to the permutation of multiple levels of multiple factors), to scenario analysis (with and without the consideration of eCO_2 and climate variability associated with climate change), and to Bayesian or probability analysis. This review aims to (1) summarise the work done so far in the area of climate change impact and adaptation in the wheat industry in Australia from both experimental and modelling perspectives, and (2) put forward future research directions to inform relevant stakeholders including policy makers and researchers.

2. EXPERIMENTAL STUDIES

Atmospheric pCO_2 and water are fundamental substances for crop to synthesize carbohydrate. Therefore the increase in atmospheric pCO_2 will have a significant implication for crop production, especially for C3 plants, by the stimulation of photosynthesis (Cure and Acock, 1986) and improvement of water use efficiency (WUE, Morison and Gifford, 1984b) and through climate change (indirect effect). The physiological effect of eCO_2 and its combined effect with temperature, solar radiation, soil water and nutrient on wheat production have been experimentally studied since 1970s by using glasshouse, gradient tunnel and free air CO_2 enrichment (FACE) facilities.

2.1. Photosynthesis, Biomass, Yield

Gifford (1977) studied the effects of eCO_2 on the phenology and growth of wheat crop under low and high light conditions. This study found that (1) there is no effect of doubled CO_2 on crop phenology, (2) most of the growth response was through tillering, (3) the growth response to eCO_2 was relatively greater under low radiation than high radiation and (4) photosynthesis increased 50% to 54%, biomass increased 8% to 36%, grain yield increased 8% to 50% with least increase in well-watered conditions, and with most increase in water-stressed canopies. Gifford (1979) examined the yield response of two wheat cultivars (Gabo and WW15) to eCO_2 (ambient plus 250 vpm) under four restricting water levels and found that there is no difference in yield response to eCO_2 across the two cultivars used and that the less water made available to the crop the greater was the response relative to the control yield. Rawson (1992) examined the interactive effects of eCO_2 and temperature on dry biomass of wheat shoot. It was revealed that wheat biomass response to eCO_2 was greater under higher temperature regime when compared with low temperature. By using temperature gradient tunnels Rawson (1995) studied the effects of doubled pCO_2 on the biomass and yield of wheat (Hartog) across summer and winter planting times. It was concluded that there was no difference in harvest index across pCO_2 levels considered while yield and biomass increased 36% and 7% for summer and winter plantings respectively.

2.2. Transpiration Efficiency and Water Use Efficiency

The reduced stomatal conductance under eCO_2 is likely to reduce water loss but maintain photosynthesis due to higher water potentials and higher internal CO_2 levels (Morison 1987) resulting in an increase in transpiration efficiency (TE). TE has been found to increase by 50-80% with doubling of CO_2 in controlled spaced plant experiments (Morison and Gifford 1984b, Samarakoon et al., 1995). However, Morison (1985) argues that the TE of an entire crop is likely to be lower due to numerous feedbacks which would occur in the field in crop canopies and he suggested that a 40% increase may be more likely. When wheat was grown as a canopy crop in glasshouse under 340 and 660ppm CO_2, TE increased by 33% under eCO_2 (Gifford and Morison, 1993). If this result is translated to a doubling of CO_2 from 350 to 700ppm it represents a 37% increase in TE. Morison and Gifford (1984a) found that doubled CO_2 increased leaf area by 67% and decreased transpiration by 43%. Morison and Gifford (1984b) found that plant dry weight increased 83-97% and water use efficiency (WUE) increased 87~94% across the two growth periods considered. Samarakoon et al (1995) compared WUE response of two wheat cultivars (Quarrion and Matong differing in stomatal conductance, (gs)) to doubled pCO_2 under wet and dry soil conditions. Findings from this study showed that cultivar (Quarrion) with lower gs has a higher CO_2 effect on WUE (+73~82%) than did Matong (+54~65%) in wet soils whereas there is a similar increase (+60 ~ 68%) in WUE in responding to eCO_2 under dry condition across the two cultivars. The reduction in stomatal conductance with increasing CO_2 level is also well documented (e.g. Morison, 1987).

2.3. Leaf N Concentration

Elevated atmospheric CO_2 levels tend to reduce leaf nitrogen concentrations of wheat (Hocking and Meyer 1991; Rogers et al., 1996) due to the change in the balance between the photosynthetic carbon reduction cycle and the photorespiratory cycle (Conroy and Hocking 1993).

Table 1. Experiments studies on the effects of eCO_2 on wheat production systems

References	Facility	Study condition	Study Components	Effects
Gifford (1977)	glasshouse	eCO_2[1]: ambient + 200 ± 20 (vpm) high and low light conditions	Phenology Growth biomass yield photosynthesis	No obvious effect via tillering, +8 ~ +36% +8 ~ +50% +50 ~ +54%
Gifford (1979)	As above	eCO_2: ambient + 250 vpm 2 cultivars 4 restricting water levels	Yield	No difference across cultivars; the less the soil water the greater the response
Samarakoon et al (1995)	As above	Doubled pCO_2[2]; 2 cultivars Wet soil dry soil	WUE	 73~82% (lower g_s cultivar) 54~65% (higher g_s cultivar) 60~68% for both cultivars
Morison and Gifford, 1984a	As above	Doubled CO_2 Soil water conditions	Leaf area Transpiration	+67% -43%
Morison and Gifford, 1984b	As above	As above	plant dry weight WUE TE	+83 ~ 97% +87 ~ 94% +50 ~ +80%
Gifford and Morison, 1993	As above	pCO_2 (ppm): 340, 660	TE	+33%
Rawson (1992)	N/A	Doubled pCO_2 (700 ppm) 2 temperature regimes	Ratio of shoot dry biomass	1.37 under 33/28°C 1.35 under 23/18°C

Table 1. (Continued)

References	Facility	Study condition	Study Components	Effects
Rawson (1995)	Gradient tunnels	Doubled pCO_2 (700 ppm) Summer plantings Winter plantings	Harvest index Yield and biomass	No difference 36% 7%
Hocking & Meyer, 1991	Glasshouse	eCO_2: 1500 cm^3/m^3 5 levels of NO_3_N supply	Stem and leaf [NO3-N] & [total-N]	substantially decreased
Rogers et al., 1996	As above	CO_2 (ppm) 350, 550, 900 N supply (mg N/kg soil): 0, 8, 17, 33, 67 & 133	Shoot growth Shoot/leaf concentration	+30% at 550 ppm across all N levels 0% at 900 ppm when N level is low Lower at each eCO_2, greatest reduction (-33%) at 900 ppm, -7% when N supply is highest ; critical foliar [N] reduced from 43 mg/g at aCO_2[3] to 39 & 38 mg/g at 550 & 900 ppm respectively
Lam et al., 2012a	AGFACE[4]	pCO_2 (ppm): 390, 550 supplementary irrigation	biomass N uptake ^{15}N recovery	27 ~ 58% at normal sowing, rainfed 18 ~ 44% at normal sowing, rainfed 20 ~ 50% from soil, no change in fertiliser N recovery irrigated vs. rainfed: 37~94% (yield); 13~609% (N uptake)`

References	Facility	Study condition	Study Components	Effects
Lam et al., 2012b	As above	pCO_2 (ppm): 390, 700 residue incorporation	Shoot, root biomass, yield, total N uptake, grain N removal and fertiliser N recovery	+16%, +41%, +19%, +4%, +8% &7% respectively without residue; positive effects of eCO_2 absent with the incorporation of residue
Taisz-Posch et al., 2012	As above	pCO_2 (ppm): 390, 550 Hartog vs. Drysdale	Grain yield	Grain yield of Drysdale is 19% higher than that of Hartog under eCO_2 and 2% higher under aCO_2.

1: enhanced CO_2; 2: CO_2 concentration; 3: ambient CO_2; 4: Australia Grain Free Air CO_2 Enrichment.

In particular, there is a consistent reduction in the critical nitrogen concentration (the concentration at which yield drops to 90% of that under optimum nutrition) that has a significant implications for wheat growth and management (Hocking and Meyer 1991, Conroy and Hocking 1993, Rogers et al., 1996).

Wheat models have been improved and modified (Reyenga et al., 1999) based on the above findings associated with the response of photosynthesis, evapotranspiration and critical nitrogen concentration to eCO_2. Ratios have been calculated between measured daily photosynthesis/evapotranspiration rates for a canopy exposed to a range of high CO_2 values, based on experimental results.

The modified wheat models such as the Agriculture Production System sIMulator (APSIM)-Iwheat/Nwheat/Wheat models were widely used to study the combined effects of eCO_2 and climate change on wheat production as described in next Section.

3. Modelling Approach

Process-oriented or mechanistic crop modelling is the most important method for climate change impact assessment and adaptation evaluation in the plant-based industries. Common used crop models in Australia include the APSIM-Wheat, APSIM-NWheat, APSIM-IWheat, the Decision Support System for Agrotechnology Transfer (DSSAT) CERES-Wheat, CropSyst etc. Normally these crop models were coupled with synthetic/arbitrary and or general circulation models (GCMs)-based climate scenarios in the below detailed sensitivity analysis, scenarios analysis and probability analysis.

3.1. Sensitivity Study

Sensitivity analysis alone was applied to many earlier agricultural impact studies and accompanied scenario analysis in later impact assessment. In this type of analysis, user modifies the historical climate conditions through an absolute difference (for temperature change) and/or a ratio change (rainfall, radiation and other variability changes). Generally, arbitrary incremental changes in climate (for example, a temperature increase of 2°C or 4°C and/or a rainfall change of ±20%) are specified and uniformly applied to the historical climate to explore the response of crop production systems to changed climate through the use of crop models (Luo and Yu, 2012). Simulation designs for sensitivity analysis have evolved from designs with a single level of a single factor, to designs with multiple levels of multiple factors, such as the permutation of changes in temperature, rainfall and pCO_2 (Luo and Kathuria, 2012). McKeon et al. (1988) found that a temperature increase of 2°C would decrease wheat yields (-6%) in Queensland, Australia. Wang et al. (1992) examined the interactive impacts of increased pCO_2 and temperature on wheat yield in Victoria, Australia and found that doubling of pCO_2 to 700 ppm would increase wheat yields by 28% to 43%, but pCO_2 increase with a simultaneous increase in temperature of 3°C would decrease yields by 25 to 60% by using current cultivars. Sensitivity study resurfaced recently due to its advantage and the fact that large uncertainties still exist in regional climate change projections including changes in inter-seasonal variation.

The combined effects of three incremental changes in temperature, rainfall and pCO_2 on Western Australian (WA) wheat production systems including production and environment across soil types were investigated in Van Ittersum et al. (2003) and Ludwig and Asseng (2006). By adopting similar approach, Luo and Kathura (2013) examined the relationship between wheat yield and climate variables in NSW. Wang et al. (2009) examined the response of wheat grain yield and biophysical processes to climate change at Wagga Wagga by conducting both sensitivity and scenario analysis.

3.2. Scenario Analysis

GCMs provide the most robust tool for constructing future climate change scenarios to be used in impact assessment and adaptation evaluation.

3.2.1. Mean Climate Change

Substantial work has been done in investigating the impacts of mean climate change on wheats yields. Reyenga et al (1999) quantified the potential impact of future climate change on wheat yield in Burnett region, Southeast Queensland under a number of climate change scenarios derived from CSIRO (1996) by using the APSIM-IWheat model. Another study by Reyenga et al. (2001) examined the impact of future climate change on wheat distribution in South Australia (SA, based on yield level). Luo et al (2003) quantified and visualised the potential impacts of future climate change including eCO_2 on wheat production in number of key wheat producing areas in SA by using DSSAT CERES-Wheat model. Using the CropSyst model, Anwar et al (2007) investigated the potential impacts of future climate change on wheat yield at Birchip, Victoria.

A most recent study by Potgieter et al. (2012) quantified the potential impacts of mean climate change on Australian wheat yields at statistical local area scale by using the OZ-Wheat model based on the outputs of CSIRO Conformal Cubic Atmospheric Model (C-CAM) driven by CSIRO Mark 3.5 GCM.

3.2.2. Changes in Mean Climate and in Climate Variability

The focus on mean climatic change has provided useful but limited information on how future changes in climatic variability (through extreme events such as drought and extreme high temperatures) might affect agriculture (Luo et al., 2010).

The importance of changes in climate variability on agricultural production has long been recognised. However, relevant studies have not been conducted until recently due mainly to a couple of reasons such as inaccessibility to daily outputs of higher resolution climate change projections, poor GCM performance in simulating the behaviour of climatic variability and lack of robust coupling techniques between downscaled GCMs outputs and dynamic crop models.

The integrated impact assessment by considering both the mean climate and climate variability is very lacking in Australia. Only one study addressed this issue. Luo et al. (2010) investigated the combined effects of changed mean climate and climate variability on the mean and coefficients of variation of grain yield and harvest index for wheat and canola crops in southeast Australia.

3.3. Probability Analysis

Probability analysis is associated with the identification, quantification and management of large uncertainties existed in this research area. Uncertainties propagate through every step of climate change impact assessment from the projection of GHGs emission, projection of climate sensitivity and projection of local climate change (climate model-to-model difference) as well as the structure and parameterisation of impact models. Two principal methods have been used to attach probabilities to impacts: (1) application of a large number of scenarios to an impact assessment and creation of a probability distribution from the outcomes; and (2) creation of an underlying probability distribution for each successive stage of analysis, explicitly managing the underlying uncertainty at each stage. Luo et al (2006), in producing a wheat yield change distribution, used the first method to manage uncertainties from the projection of GHG emission, projection of climate sensitivity and projection of regional climate change.

The second method is described by Hulme and Carter (1999) and Jones (2000) and applied in a number of studies in Australia. Howden et al. (1999) investigated the potential impacts of climate change on wheat yield, grain nitrogen content and gross margin at nine locations in the wheat grain belt of Australia for the year 2100, by adopting a probabilistic approach with pCO_2 fixed at 700 ppm. Howden and Jones (2001) undertook a probabilistic analysis on the costs and benefits of climate change impact on Australia's national wheat crop. Luo et al. (2005a, b, c, 2007) extended the work of Howden et al (1999) by considering the uncertainties of projected atmospheric pCO_2 in addition to the uncertainties of projected temperature and rainfall with emphases on the construction of probabilistic climate change scenarios, probabilistic yields, spatial variability of climate change impact and farm profitability impact.

3.4. Adaptation Evaluation

Climate risk assessment includes two important components: impact assessment and adaptation evaluation. Some of impact studies reviewed above have adaptation evaluation component by considering a range of planting times, varieties (Wang et al., 1992; Howden et al., 1999; Reyenga et al.,1999), a number of production regions/locations, soil types and/or nitrogen application rates (Van Ittersum et al., 2003; Ludwig and Asseng, 2006; Luo and Kathuria, 2013). Climate change adaptation has become a hot topic in nowadays policy, research and business. Luo et al. (2009) assessed the impact of management options in mitigating climate change risk such as changing sowing time, fertiliser application rate and wheat cultivars at Keith, SA and found that these options are very limited in maintaining existing wheat crop yield unless irrigation practice is introduced which is limited by the availability of irrigation water. A wide range of research activities are underway in this very important industry in an era of promoting adaptation actions.

Table 2. Modelling the impact of climate change and management options on wheat grain yield in Australia

Studies	Study Areas	Yield Change (%)	Crop Model	Climate Change Scenarios	eCO_2 Effect	Grain Nitrogen Content (GNC)	Climate Variability	Adaptation options	Market or Economic Issue
McKeon et al., 1988	Queensland (Qld)	-6 +23	N/A	+2°C T[1] increase + R^2 change	No				
Wang et al., 1992	Victoria	+28 to +43 -60 to –25 increase	N/A	CO_2 alone CO_2 + T (3°C) CO_2 +T (3°C) + late maturity variety	yes			Change cultivars	
Reyenga et al., 1999	Southeast Qld	+26 to +37 +9 to +23 +21 to +33 -3 to +6 +6 to +26	I_Wheat	CO_2 alone CO_2 + T (+2.76 °C) CO_2+ T(+2.76 °C) + R (+) CO_2+ T(+2.76°C) +R (-) CO_2 + climate change + fertilizer application	yes	-12 to –11 -9 to –8		Changing N application rates	
Reyenga et al., 2001	South Australia (SA)	+13 to +52 -35 to -10	I_Wheat	CO_2 alone CO_2 + dry scenarios	yes				
Howden et al., 1999	Australia wide	+5 to +43 +9 to +37 +5 to 46	I_Wheat	CO_2 alone CO_2 + Climate Change CO_2 + Climate Change +changing planting window	yes	-14 to -4		Change planting time,	+6 to +70 +5 to +60 +28 to +95 (gross margin)

Studies	Study Areas	Yield Change (%)	Crop Model	Climate Change Scenarios	eCO_2 Effect	Grain Nitrogen Content (GNC)	Climate Variability	Adaptation options	Market or Economic Issue
Luo et al., 2003	SA	Effects of eCO_2 offset the negative effects of T & R	DSSAT CERES-Wheat, v3.5	Seasonal changes in T & R	yes	Decreased under all pCO_2 and most CCSs			
Luo et al., 2005a	Roseworthy, SA		The APSIM-Wheat	Outputs of nine downscaled GCMs from OZClim: probabilistic scenarios	yes				Risk increased 18% under the most likely scenario
Luo et al., 2005b	SA	-32% ~ -13.5% across all locations	As above	As above under most likely scenario	yes				
Luo et al., 2005c	Lower North, SA	-58%~-3% (most likely scenario)	As above	The best, median and worst scenarios	yes	−25% ~ +42% across all scenarios			
Luo et al., 2006	SA	negatively impacted by climatic change with 80~96% chance	Statistical modelling	Outputs of nine downscaled GCMs from OZClim: probabilistic scenarios	yes				
Luo et al., 2007	SA		Bio-economic modelling	Most likely CCSs[3]	yes				Risk increased more or less across all locations

Table 2. (Continued)

Studies	Study Areas	Yield Change (%)	Crop Model	Climate Change Scenarios	eCO_2 Effect	Grain Nitrogen Content (GNC)	Climate Variability	Adaptation options	Market or Economic Issue
Luo et al., 2009	Keith, SA	Chosen adaptation options cannot offset the negative effects	The APSIM-Wheat	Outputs of CSIRO Conform Cubic Atmospheric Model - dynamically downscaled data	yes		yes	Change planting time, cultivars and N application rates	
Luo et al., 2010	Southeast Australia	Changes in CV[4] has negative effects on wheat yields	The APSIM-Wheat	As above	yes		yes		
Luo and Kathuria, 2013	NSW	options preference: Soil> site > changes in pCO_2 > R > T	The APSIM-Wheat	Incremental changes in R, T & pCO_2	yes			Soil types, locations	
Wang et al., 2009	Wagga Wagga, NSW	Increased 1% in 2050 and -6% in 2070	As above	Incremental change of R, T & pCO_2 and outputs of MK2 via OZClim					
Anwar et al., 2007	Birchip, Victoria	-25%	CropSyst v4	Monthly outputs of CCAM driven by CSIRO Mark 2 & 3	yes				
Ludwig and Asseng (2006)	Western Australia	Negative effects due to 15% reduction in R can be compensated by the increase of T (2°C) and of pCO_2 (50%)	APSIM-Nwheat	Incremental change of R, T & pCO_2	yes	eCO_2 reduce GNC; lower R increased GNC; higher T increase or decrease GNC		Locations, soil types	

Studies	Study Areas	Yield Change (%)	Crop Model	Climate Change Scenarios	eCO_2 Effect	Grain Nitrogen Content (GNC)	Climate Variability	Adaptation options	Market or Economic Issue
Van Ittersum et al., 2003	As above	Yield increased under eCO_2, high N input and drier conditions; non-linear relationship with high T (3°C),	As above	As above	yes	Increased pCO_2 and T decreased GNC		Locations, soil types, N application rates, and varieties	
Potgieter et al., 2012	Australia	2020: minor change 2050: -30% ~ +6%	OZ-Wheat	Monthly outputs of CCAM driven by CSIRO Mark 3.5 under B1 and A1T	yes				

1: temperature, 2: rainfall, 3: climate change scenarios, 4: climate variability.

4. Emerging Researches and Future Research Needs

FACE is recognised as the most realistic facility in examining the direct effects of eCO_2 on plant/crop responses because plants are not enclosed. Department of Climate Change and Energy Efficiency commissioned the establishment of Australian Grain Free Air CO_2 Enrichment (AGFACE) facilities in the northwest (Horsham and Walpeup) of Victoria in 2006. AGFACE was designed to investigate the interactive effects of eCO_2 with a range of environmental factors (i.e. soil water and nutrients, temperature regimes as represented by different sowing times) on wheat production (Mollah et al., 2012). Tausz-Posch et al (2012) compared the performance of two wheat cultivars (Drysdale and Hartog) in a high CO_2 environment and found that Drysdale with higher TE performed better than Hartog in terms of grain yield under eCO_2.

Recently, the AGFACE has been extended to study the effects of eCO_2 on belowground processes of a wheat-pulse rotation system across three major soil types at Horsham (Jin et al., 2012). Lam et al. (2012a) studied the effects of eCO_2 on growth, N uptake and N^{15} recovery and found that (1) eCO_2 did not alter the recovery of fertiliser N in plant and in soil, (2) grain yield and whole plant N uptake under supplementary irrigated plots were 37~94% and 13~609% respectively higher than the rainfed counterparts. Lam et al (2012b) investigated the interaction between crop residue incorporation and eCO_2 on the growth, grain yield and the recovery of ^{15}N-labeled fertiliser in the wheat production system. Experimental results showed that eCO_2 increased wheat shoot biomass (16%), root biomass (41%), grain yield (19%), total N uptake (4%), grain N removal (8%) and fertiliser N recovery in the plant (7%) without the incorporation of residue whereas the positive effects of eCO_2 was absent when residue was incorporated. More information on these studies can be found in Table 1.

Crop models played an important role in transferring climate change information including eCO_2 to yield impact. Therefore the robustness of crop models is very important for reliable impact assessment and the identification of effective adaptation options. Crop model itself including model structure and parameters has been identified as one of the many sources of uncertainties in the area of climate change impact and adaptation in the agricultural sector (Luo et al., 2005a). Even though those crop models used in Australia have been validated and/or modified to represent the direct effects of eCO_2 on wheat production, the validation and modification processes were based on earlier domestic glasshouse experimental studies and/or overseas experimental studies. As mentioned earlier, FACE provides the most realistic facility to quantify the effects of eCO_2 along with other environmental factors on crop production. Overseas experimental studies including FACE with less limiting soil water and nutrients conditions, used to validate and/or modify those crop models applied in Australia, do not reflect Australian wheat germplasm and do not represent existing and future Australian environment (lower soil fertility and drier environment). The AGFACE described above provides invaluable datasets relevant to Australian wheat genetic resources and future crop growing conditions to improve the robustness of crop models and the accuracy of the resultant climate change impact assessment and adaptation evaluation in the wheat industry. With state-of-the-art AGFACE datasets available the logic step next is to validate and improve the most commonly used wheat models, a prerequisite for confident impact assessment and adaptation evaluation.

In addition to validation and improvement of these wheat models, inter-comparison of these wheat models is also needed to identify the best performance model to be applied to climate change impact and adaptation studies. This type of work is lack globally in general and in Australia in particular. A couple of studies addressed uncertainty issues from the use of different crop growth models. Wolf et al. (1996) validated and calibrated five commonly used wheat models and applied them to quantify climate change impact on wheat production in Europe. Ewert et al. (2002) tested the performance of three wheat crop models with different complexity against experimental datasets from FACE and Open Top Chamber. There is an increasing interest in inter-comparing climate change impact and adaptation results by adopting different crop models and/or with different complexity associated with agricultural risk assessment, for example, the global Agricultural Model Inter-comparison and Improvement Project (www.agmip.org). Inter-crop model comparison is urgent needed in Australia to reach robust conclusions regarding the magnitude of climate change impacts and the effectiveness of on-farm adaptation and mitigation strategies.

REFERENCES

ABARE 2008, Australian Commodities March Quarter, vol. 8, no. 1, ABARE, Canberra.

Anwar, M.R., O'Leary, G., McNeil, D., Hossain, H. and Nelson, R.: 2007, Climate change impact on rainfed wheat in south-eastern Australia. *Field Crops Research,* 104 139–147.

Conroy, J. and Hocking, P. J.: 1993, Nitrogen Nutrition of C3 Plants at Elevated Atmospheric CO_2 Concentrations. *Physiologia Plantarum,* 89 570-576.

CSIRO 1996 scenarios http:www.dar.csiro.au/pub/programs/climod/cm4.htm.

Cure, J. D. and Acock, A.: 1986, Crop Response to Carbon Dioxide Doubling: A Literature Survey. *Agricultural and Forest Meteorology,* 38, 127-145.

Ewert, F., Rodriguez, D., Jamieson, P., Semenov, M.A., Mitchell, R.A.C., Goudriaan, J., Porter, J.R, Kimball, B.A., Pinter Jr, P.J., Manderscheid, R., Weigel, H.J., Fangmeier, A., Fereres, E. and Villalobos, F.: 2002, Effects of elevated CO_2 and drought on wheat: testing crop simulation models for different experimental and climatic conditions. *Agriculture, Ecosystems and Environment,* 93 249-266.

Gifford, R. M.: 1977, Growth Pattern, Carbon Dioxide Exchange and Dry Weight Distribution in Wheat Growing under Differing Photosynthetic Environments. *Aust. J. Plant Physiol.,* 4, 99-110.

Gifford RM (1979) Growth and Yield of CO2-Enriched Wheat Under Water-Limited Conditions. *Australian Journal of Plant Physiology,* 6, 367–378.

Gifford, R. M. and Morison, J. I. L.: 1993, Crop Response to the Global Increase in Atmospheric Carbon Dioxide Concentration. *In International Crop. Science I, CSSA, USA,* 325-331.

Hocking, P. J. and Meyer, C. P.: 1991, Carbon Dioxide Enrichment Decreases Critical Nitrate and Nitrogen Concentrations in Wheat. *J. Plant Nutrition,* 14 (6) 571-584.

Howden, S. M., Reyenga, P. J. and Meinke, H.: 1999, Global Change Impacts on Australian Wheat Cropping. Report to the Australian Greenhouse Office.

Howden, S. M. and Jones, R. N.: 2001, Cost and benefits of CO2 increase and climate change on the Australian wheat industry. Australian Greenhouse Office, Canberra; www.greenhouse.gov.au/science/wheat/index.html.

Hulme, M. and Carter, TR.: 1999, Representing Uncertainty in Climate Change Scenarios and Impacts Studies. In: Representing Uncertainty in Climate Change Scenarios and Impacts Studies. [Carter, TR., Hulme, M., Viner, D. (eds.)]. Proceedings of the ECLAT-2 Helsinki Workshop, 14-16 April, 1999, Climatic Research Unit, Norwich, p11-37.

Jin J, Tang C, Armstrong R, Sale P (2012). Phosphorus supply enhances the response of legumes to elevated CO_2 (FACE) in a phosphorus-deficient Vertisol. Plant and Soil, in press.

Jones, R.N.: 2000, Analyzing the Risk of Climate Change Using an Irrigation Demand Model. *Climate Research,* 14:89-100.

Lam SK, Chen D, Norton R, Armstrong R (2012a). Crop residue incorporation negates the positive effect of elevated atmospheric carbon dioxide concentration on wheat productivity and fertilizer nitrogen recovery. *Plant Soil*, in press. DOI 10.1007/s11104-012-1448-4.

Lam SK, Chen D, Norton R, Armstrong R (2012b). Nitrogen demand and the recovery of 15N-labelled fertilizer in wheat grown under elevated carbon dioxide in southern Australia. Nutrient Cycling in Agroecosystems, in press. DOI: 10.1007/s10705-011-9477-6.

Ludwig F, Senthold A (2006) Climate change impacts on wheat production in a Mediterranean environment in Western Australia. *Agr. Syst.,* 90:159–179.

Luo Q, Williams M, Bellotti WD, Bryan B (2003) Quantitative and visual assessment of climate change impacts on South Australian wheat production. *Agric Syst.,* 77:173–186.

Luo Q, Bellotti WD, Williams M, Bryan B (2005a) Potential impact of climate change on wheat yield in South Australia. *Agr. Forest Meteorol.,* 132(3–4):273–285.

Luo Q, Bryan B, Bellotti WD, Williams M (2005b) Spatial analysis of environmental change impacts on wheat production in mid-lower north, South Australia. *Clim. Chang.,* 72(1–2):213–228.

Luo Q, Jones R, Williams M, Bryan B, Bellotti WD (2005c) Construction of probabilistic distributions of regional climate change and their application in the risk analysis of wheat production. *Clim. Res.,* 29(1):41–52.

Luo Q, Bellotti WD,Williams M, Jones R (2006) Probabilistic analysis of potential impacts of climate change on wheat production in contrasting environments of South Australia. *World Resour. Rev.,* 17(4):517–534.

Luo Q, Bellotti WD, Williams M, Cooper I, Bryan B (2007) Risk analysis of possible impacts of climate change on South Australian wheat production. *Clim. Chang.,* 85:89–101.

Luo Q, Bellotti WD, Williams M, Wang E (2009) Adaptation to climate change of wheat growing in South Australia: analysis of management and breeding strategies. *Agric. Ecosyst. Environ.,* 129:261–267.

Luo Q, Bellotti WD, Hayman P, Williams M, Devoil P (2010) Effects of changes in climatic variability on agricultural production. *Clim. Res.,* 42:111–117.

Luo, Q. and Yu, Q, 2012, Developing Higher Resolution Climate Change Scenarios for Agricultural Risk Assessment: Progress, Challenges and Prospects. *International Journal of Biometeorology,* 56 (4) 557-568.

Luo, Q and Kathuria, A.: 2013, Modelling the Response of Wheat Grain Yield to Climate Change: A Sensitivity Analysis. *Theoretical and Applied Climatology* **111** (1-2) 173-182.

Mckeon, G. M., Howden, S. M., Silburn, D. M., Carter, J. O., Clewett, J. F., Hammer, G. L., Johnston, P.W., Lloyd, P. L., Mott, J. J., Walker, B., Weston, E. J. and Wilcocks, J. R.: 1988, The effect of Climate Change on Crop and Pastoral Production in Queesland. In: Pearman, G. I. (ed) Greenhouse: Planning for Climate Change. CSIRO, Melbourne, Australia, 546-563.

Mollah M, Partington D, Fitzgerald G (2012). Understand distribution of carbon dioxide to interpret crop growth data: Australian grain free air carbon dioxide enrichment experiment. *Crop and Pasture Science*, Vol 62 (10), 883-891. http://dx.doi.org/10.1071/CP11178.

Morison, J. I. L. and Gifford, R. M.: 1984a, Plant Growth and Water Use with Limited Water Supply in High CO_2 Concentrations. I. Leaf Area, Water Use and Transpiration. *Aust. J. Plant Physiol.,* 11, 361-74.

Morison, J. I. L. and Gifford, R. M.: 1984b, Plant Growth and Water Use with Limited Water Supply in High CO_2 Concentrations. II. Plant Dry Weight, Partitioning and Water Use Efficiency. *Australian Journal of Plant Physiology,* 11, 375-384.

Morison, J. I. L.: 1985, Sensitivity of Stomata and Water Use Efficiency to High CO_2. *Plant, Cell and Environment,* 8, 467-474.

Morison, J. I. L.: 1985, Sensitivity of Stomata and Water Use Efficiency to High CO_2. *Plant, Cell and Environment,* 8, 467-474.

Morison, J. I. L.: 1987, Intercellular CO_2 Concentration and Stomatal Response to CO_2. In Stomatal Function, eds E. Zeiger, G. D. Farquhar and I. R. Cowan. Stanford University Press, Stanford, California, USE, 229-251.

Potgieter, A., Meinke, H., Doherty, A., Sadras, V.O., Hammer, G., Crimp, S. and Rodriguez, D.: 2012, Spatial impact of projected changes in rainfall and temperature on wheat yields in Australia. *Climatic Change,* 113 (2) pp.

Rawson, H. M.: 1992, Plant Responses to Temperature under Conditions of Elevated CO_2. *Aust. J. Bot.,* 40 473-90.

Rawson, H. M.: 1995, Yield Responses of Two Wheat Genotypes to Carbon Dioxide and Temperature in Field Studies using Temperature Gradient Tunnels. *Aust. J. Plant Physiol.,* 22 23-32.

Reyenga, P. J., Howden, S. M., Meinke, H. and McKeon, G. M.: 1999, Modelling global change impacts on wheat cropping in south-east Queensland, Australia. *Environmental Modelling and Software,* 14 297–306.

Reyenga PJ, Howden SM, Meinke H, Hall WB (2001) Global change impacts on wheat production along an environmental gradient in South Australia. *Enviro Int.,* 27:195–200.

Rogers, G., Milham, P. J., Gillings, M. and Conroy, J. P.: 1996, Sink Strength May Be the Key to Growth and Nitrogen Response in N-Deficient Wheat at Elevated CO2. *Australian Journal of Plant Physiology,* 23 253-264.

Samarakoon, A. B., Müller, W. J. and Gifford, R. M.: 1995, Transpiration and Leaf Area under Elevated CO2: Effects on Soil Water Status and Genotype in Wheat. *Australian Journal of Plant Physiology,* 22, 33-44.

Tausz-Posch S, Seneweera S, Norton RM, Fitzgerald G, Tausz M (2012). Growth and yield performance of two wheat cultivars contrasting in transpiration efficiency under elevated CO2. Field Crops Research, in press.

Van Ittersum MK, Howden SM, Asseng S (2003) Sensitivity of productivity and deep drainage of wheat cropping systems in a Mediterranean environment to changes in CO2, temperature and precipitation. *Agric. Ecosys. Env.,* 97:255–273.

Wang YP, Handoko J, Rimmington GM (1992) Sensitivity of wheat growth to increased air temperature for different scenarios of ambient CO_2 concentration and rainfall in Victoria, Australia—a simulation study. *Clim. Res.,* 2:131–149.

Wang J, Wang E, Luo Q, Kirby M (2009) Modelling the sensitivity of wheat growth and water balance to climate change in Southeast Australia. *Clim. Chang.,* 96:1–2.

Wolf, J., Evans, L.G., Semenov, M.A., Eckersten, H. and Iglesias, A.: 1996, Comparison of wheat simulation models under climate change. I. Model calibration and sensitivity analyses. *Clim. Res.,* **7** 253-270.

In: Crop Yields
Editors: Lijuan Huang and Qiao Zhao
ISBN: 978-1-62618-175-5

Chapter 4

IMPROVEMENT OF COTTON PRODUCTION IN ARID SALINE SOILS BY BENEFICIAL MICROBES

Egamberdieva Dilfuza and Jabborova Dilfuza
Faculty of Biology and Soil Sciences, National University of Uzbekistan, Vuzgorodok, Tashkent, Uzbekistan

ABSTRACT

Crop cultivation in saline soils is one of the major agricultural challenges world-wide. Cotton production in Uzbekistan is limited by soil salinization and disease caused by soil borne pathogens. The use of plant growth promoting rhizobacteria (PGPR) to control plant diseases and stimulation of plant growth has been considered a viable alternative and environmentally friendly method. The strain *P. putida* BIST and *B. subtilis* SUBTIN significantly ($p<0.05$) increased the total N, P, and K contents of shoot and root (13-27%), plant height (15%), and yield (16%) of cotton in small plot experiments. They were able to produce phytohormone indole-acetic-acid (IAA) under saline conditions and were positive for ACC deaminase activity. On the basis of results, it may be concluded that plant growth promoting rhizobacteria are potential option for improvement of cotton growth and development in salinated arid soil.

INTRODUCTION

Salinity and drought are the main causes of desertification which strongly influence many properties and processes of living organisms. Salinity causes a disturbance of plant-microbe interaction which is a critical ecological factor to help further plant growth in degraded ecosystems (Requena et al., 2001; Egamberdiyeva et al., 2007). As a result of soil salinization, plants are under saline or water unbalance stress and become more vulnerable to diseases, often caused by pathogenic fungi which can hardly be overcome by conventional methods of pest management (Werner and Finkelstein, 1995). Uzbekistan, located in Central Asia, is an example of a country, in which the soil salinity is a major concern. In 1990, about 48% of the total irrigated land area was suffering from the soil salinity, and by 2000, the area of salinized soil covered already 64% of the irrigated land area (Shirokova et al., 2000; Egamberdiyeva et al., 2007). Cotton (*Gossypium hirsutum*) is the largest agricultural crop in Uzbekistan. Its annual production is about 3.5-4 million tons (2007), making Uzbekistan the world's sixth largest producer and second largest exporter of cotton (UzStat 2007). Cotton production in Uzbekistan is limited by soil salinization and disease caused by soil borne pathogens such as *Fusarium, Rhizoctonia* and *Verticillium* (El-Sayed and Embaby 2007). In Uzbekistan, estimated annual losses caused by both the soil salinity and fungus-caused root diseases are estimated to be 30% (Sheraliev et al., 2008). The chemicals used to protect plants against fungal pathogens are harmful to the environment. Under such circumstances, it requires suitable biotechnology not only to improve crop productivity but also to improve soil health through interactions of plant roots and soil microorganisms (Lugtenberg and Kamilova, 2004). The use of plant growth promoting rhizobacteria (PGPR) to control plant diseases and stimulation of plant growth has been considered a viable alternative and environmentally friendly method (Mayak et al., 2004; Egamberdieva et al., 2008; Lugtenberg and Kamilova, 2009).

Bacteria that exert beneficial effects on plant development known as plant growth-promoting rhizobacteria (PGPR) have been reported widely. The stimulatory effect of PGPR on the seed germination, plant growth and nutrient uptake of various plants were well addressed by Lugtenberg et al.(2001), Glick et al.(2007), Adesemoye and Kloepper(2009) and Egamberdieva and Kucharova (2009). They can also enhance plant's tolerance to adverse environmental stresses, which include salt stress (Egamberdieva, 2008; Maheshwari et al., 2011), drought stress (Zahir et al., 2008), and heavy metal contaminations (Sheng, 2005). A range of salt tolerant rhizobacteria (e.g.,

Rhizobium, *Azospirillum*, *Pseudomonas*, *Flavobacterium*, *Arthrobacter* and *Bacillus*) has shown beneficial interactions with plants in stressed environments (Lugtenberg et al., 2001; Egamberdieva and Islam, 2008; Arora et al., 2008). They may also enhance mineral uptake, not only as a consequence of the increase in root surface area, take up N, P, K, and microelements more efficiently from the soil but also by stimulating the ion uptake systems (Burdman et al., 2000).

Mechanisms by which bacteria can promote plant growth and protect plants from phytopathogens include mobilization of nutrients (Lugtenberg and Kamilova, 2004), production of phytohormones such as indole acetic acid (IAA), gibberellins (Frankenberger and Arshad, 1995; Egamberdieva et al., 2010), hydrogen cyanide (HCN) (Duffy et al., 2004), antibiotics (Chakraborty et al., 2009) and synthesis of the enzyme 1-amino cyclopropane-1-carboxylate (ACC) deaminase (Patten and Glick, 2002). Many studies have shown that microbial isolates can effectively control *R. solani*-induced damping off of cotton seedlings both in the laboratory and field conditions (Hassanin et al., 2007; Gasoni and Gurfinkel, 2009).

Application of bacterial inoculants such as *Azotobacter, Azospirillum, Pseudomonas*, has improved growth, nutrient uptake and yield of cotton in various climatic conditions (Anjum et al., 2007; Hafeez et al., 2002; Rashid et al., 2000). However, most studies on improvement of growth, development and yield of cotton have been conducted in non-saline agricultural soils and have not addressed the problems associated with the salinity. The objective of the research was to determine the ability of rhizobacteria to enhance the growth, nutrient uptake (N,P,K) and yield of cotton grown in salinated arid soil of Uzbekistan.

Material and Methods

Study Site and Soil Characterisation

The study site located in Syrdarya province (41° 00′ N, 64° 00′ E,) in north-eastern Uzbekistan. According to the WRB-FAO (2006) classification, the soils of selected fields were identified as Calcisol (silt loam seirozem). The surface soil horizon was calcareous saline whereas the deeper soil horizons were only mildly alkaline (Egamberdiyeva et al., 2007). In these soils, cotton has been grown for the last 50 to 60 years under a continuous monoculture production system and under flood irrigation without proper drainage facilities but using a natural flow system. The climate of the area is continental with a yearly average rainfall of 200 mm and more than 90 percent of the total rain falling from October to May. The average minimum monthly air temperature is 0°C in January and the maximum of 37°C is in July, while the soil temperature ranges from -2 to +35°C. The average highest relative humidity is slightly more than 80% in January and minimum of less than 45% in June. Under a dry continental climate, the combination of high temperature and low rainfall during the growth season makes irrigation essential for crop production. In general, high concentration of Ca^{2+}, K^{+}, and Na^{+} are associated with CO_3^{2-} and Cl^{-} ions, reflecting the dominance of carbonates and chlorides in saline soil. On average, the two kinds of soils was taken contained 42 ± 9 g of sand kg^{-1}, 708 ± 12 g of silt kg^{-1}, and 250 ± 13 g clay kg^{-1} (Egamberdieva and Kucharova 2009). Electrical conductivity (EC) values of saline soil were 7.2 dS/m. The main chemical soil properties are: organic matter 0.79 %; Ct 2.39 %; Nt 0.07 %; CO_3^{2-}-C 1.59%, Ca^{2+} 54.3 g/kg; Mg^{2+} 26.1 g/kg; K^{+} 6.7 g/kg; P 1.2 g/kg; Cl^{-} 0.1 g/kg; Na^{+} 0.8 g/kg; pH 8.0. The high concentration of Ca^{2+}, K^{+}, and Na^{+} are associated with CO_3^{2-} and Cl^{-} reflecting the dominance of carbonate and chloride in saline soil.

Plant and Microorganisms

Seeds of the salt-tolerant cotton cultivar 'Namangan' were obtained from the Tashkent State University of Agriculture, Faculty of Plant Production. All bacterial isolates were obtained from the Culture Collection of the National University of Uzbekistan (CCNUU). The strains *P. putida* BIST, *P. stutzeri* NUU8 and *B. subtilis* SUBTIN were previously isolated from the rhizosphere of cotton grown in salinated Uzbek soil. *Pseudomonas* strains were grown overnight in KB broth King's B medium (KB), whereas *Bacillus* was grown in Luria-Bertani broth (LB) (Difco Laboratories). 1.0 ml of an overnight culture was pelleted (13,000 × *g*) and the supernatant was discarded. Cells were washed with 1 ml of PBS and re-suspended (Leeman et al., 1995). Cell suspensions were adjusted to OD_{620} = 0.1 for *Pseudomonas* and OD_{620} = 0.3 for other bacterial strains. The both of OD_{620} values corresponded to a cell density of 10^7 - 10^8 cells/ml and were used for inoculation of sterile cotton seedlings.

Plant Growth Promotion by Bacteria

The effect of the bacterial strains on the growth of cotton was measured in plastic pots containing 300 g of the saline soil mentioned above. The inoculation treatments were set-up in a randomised design with 10 replications. After six weeks of growth the root and shoot length and dry weight of the whole plants was determined.

The field experiments were conducted using a randomized complete block design (RCBD) described by Clewer and Scarisbrick (2001) at the experimental station. The site of experiment was ploughed to 0.25 m depth after the removal of the previous crop. The biohumus served as carriers for the **bacteria** and bacterial formulation was prepared as described above, where bacterial density CFU 10^7/g biohumus. The treatments were: control, *P. putida* BIST, *P. stutzeri* NUU8 and *B. subtilis* SUBTIN. The cotton seeds were inoculated (mixed) with bacterial formulation and were sown by hand in 7 m x 4 m plots. Intra-row spacing was 20 cm and rows were 60 cm apart and there were four replicate plots per treatment (total 12 treatments) At harvest time, cotton plants were evaluated for plant height, yield per plant and biological yield.

Nutrient Uptake

For determination of nutrient uptake of root and shoot, the plants separated to root and shoot and were dried in 105°C. Total N, P contents were determined by using a CNS elemental analyzer (LECO® Corporation, St. Joseph, MI, USA) according to DIN ISO 15178 (2001) and Potassium, K, was analyzed according to DIN 38414-S (1983).

1-Aminocyclopropane-1-Carboxylic Acid (ACC) Deaminase Activity

1-Amino cyclopropane-1-carboxylic acid (ACC) deaminase is an enzyme that degrades the precursor of the plant hormone ethylene, which is produced by the plant during environmental stress. Some PGPR bacteria are able to promote plant growth by lowering the endogenous ethylene synthesis in the roots through their ACC deaminase activity (Glick *et al.* 1998). The method of analysing bacterial strains for their ability to use ACC as the sole nitrogen source is a trait that is a consequence of the presence of the activity of the enzyme, ACC deaminase. Synthetic Basal Medium (BM) (Lugtenberg et al.,

1999) was used to check whether bacterial strains had ACC deaminase activity. BM medium was supplemented with 3.0 mM ACC (Sigma Chemical Co., St. Louis, MI, USA) (**EC** 4.1.99.4) and 3.0 mM $(NH_4)_2SO_4$ as the sole N source (positive control) or without an added N-source (negative control).

Production of Indole-Acetic Acid (IAA)

Indole-3-acetic acid (IAA) is the most abundant endogenous auxin produced by plants, and which regulates aspects such as stem elongation and root growth. The production of IAA by bacteria is as an important factor in direct plant-growth-promoting abilities of rhizosphere bacteria (Frankenberger and Arshad, 1995; Woodward and Bartel, 2005). The production of IAA was determined according to the method of Bano and Musarrat (2003). *Pseudomonas* strains were grown in KB medium and *Bacillus* strain were grown in Luria–Bertani (LB)medium supplemented with 1.5% NaCl with and without tryptophan (500 µl/ml) and incubated at 28°C. After three days of cultivation, 2 ml aliquots of bacterial cultures were centrifuged at 13,000 × g for 10 min. One ml of supernatant was transferred to a fresh tube to which 100 µl of 10 mM orthophosphoric acid and 2 ml of reagent (1 ml of 0.5 M $FeCl_3$ in 50 ml of 35% $HClO_4$) were added. After 25 min, the absorbance of the developed pink color was measured at 530 nm using a Perkin-Elmer **Lambda** 3A **spectrophotometer** (**Perkin-Elmer** Corp., Norwalk, CT, USA). The concentration of IAA in culture was calculated from a calibration curve of pure IAA standard.

Statistical Procedures

Data were tested for statistical significance using the analysis of variance package included in Microsoft Excel 2007. Comparisons were done using Student's *t*-test. Mean comparisons were conducted using a least significant difference (LSD) test ($P < 0.05$).

RESULTS

In our study the PGPR strains *B. subtilis* SUBTIN, *P. putida* BIST and *P. stutzeri* NUU8 stimulated shoot and root dry matter of cotton in salinated soil (Table 1). The untreated plants showed very poor growth in nutrient deficient soil, where high pH make nutrients less available to plants. The inoculation of cotton with bacteria *P. putida* BIST and *B. subtilis* SUBTIN significantly ($p<0.05$) increased the shoot dry weight from 14 to 24% as compared to the control plant (Table 1). Only strain *P. stutzeri* NUU8 was not effective for shoot and root growth of cotton.

Table 1. The effect of bacteria strains on root and shoot dry matter of cotton in salinated soil (pot experiment, control=0.310 g/shoot, 0.185 g/root (100%)

Bacterial Strains	Shoot	Root
Control	100 (0.310)	100 (0.185)
P. stutzeri NUU8	110	113
P. putida BIST	114*	114*
B. subtilis SUBTIN	124*	119*
LSD $\alpha = 0.05$	11	14

* Significantly different from the control at P<0.05

Table 2. The effect of bacteria strains on N,P,K uptake of cotton shoot and root in salinated soil (pot experiment, control (N) = 0.0048 μg/g shoot, 0.0019 μg/g root; (P) 0.00042 μg/g shoot, 0.00026 μg/g root; (K) 0.0026 μg/g shoot, 0.0021 μg/g/ root (100%)

Bacterial strains	N		P		K	
	Shoot	Root	Shoot	Root	Shoot	Root
Control	100 (0.0048)[1]	100 (0.0019)	100 (0.00042)	100 (0.00026)	100 (0.0026)	100 (0.0021)11
P. stutzeri NUU8	106	110	110	111	112	8*
P. putida BIST	113*	106	123*	109	126*	113
B. subtilis SUBTIN	111	103	116*	120*	127*	121*
LSD $\alpha = 0.05$	13	17	14	17	18	18

* Significantly different from the control at P<0.05

The increase in plant biomass translated into significantly higher N, P, and K contents. The strain *P. putida* BIST and *B. subtilis* SUBTIN significantly ($p<0.05$) increased (13-27%) the total N, P, and K contents of shoot and root

(Table 2). The most effective strain was *B. subtilis* SUBTIN which generated 27% increase in K uptake of shoot over the control plant (Table 2). The P and K uptake of shoot and root was increased more than N uptake. Only strain *P. putida* BIST increased the N - uptake by shoot.

Our results showed that bacterial inoculants *P. putida* BIST and *B. subtilis* SUBTIN increased plant height, biological yield and yield of cotton in small plot experiments. The plant height was stimulated after inoculation with *B. subtilis* SUBTIN up to 15% compared to un-inoculated control plants. *P. putida* BIST increased cotton yield by 16 % and biological yield up to 10% (Table 3).

Table 3. The effect of bacteria inoculants on cotton growth and yield in small field experiments

Treatments	Plant hight cm	%	Yield (g plant $^{-1}$)	%	Biological yield (g plant $^{-1}$)	%
Control	110.2	100	40.8	100	280.2	100
P. putida BIST	123.1	111	47.5*	116*	310.7*	110
B. subtilis SUBTIN	127.4*	115	45.7	112	300.9	107

* Significantly different from the control at P<0.05

The production of auxin (IAA) of the strains *P. putida* BIST and *B. subtilis* SUBTIN was tested by growing in KB medium under saline condition (1-4% NaCl). Both strains tested appeared to produce IAA in media containing up to 3% NaCl (Figure). The presence of tryptophan stimulated auxin production in *P. putida* BIST and *B. subtilis* SUBTIN.

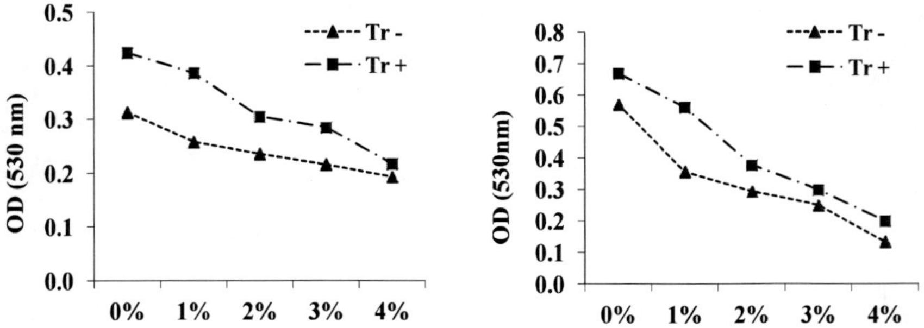

Figure. The production of IAA (indole-3-acetic-acid) by *P. putida* BIST (A) and *B. subtilis* SUBTIN (B) under various NaCl conditions (bacteria were grown 6 days at 28°C with the absence and presence of 100 ug/ml of the auxin precursor tryptophan).

Only strain *P. putida* BIST could utilize ACC as the sole N source, indicating the presence of ACC deaminase, which plays a role in reducing the levels of the stress compound ethylene in plants. Other strains were negative the ACC deaminase activity.

Discussion

In salinated soil two strains significantly ($P < 0.05$) increased root and shoot dry weight, and yield of cotton. Similar results were reported by Yue et al. (2007) for *Klebsiella oxytoca*, which was able to relieve salt stress and promote the growth of cotton seedlings in salinated soil. After treatment with bacterial strains, plant height and dry weight of cotton increased by 14.9 and 26.9%, respectively, compared to the control. In K-deficient soil, *Bacillus edaphicus* also stimulated growth of cotton and the N and P content of above-ground plant components (Sheng, 2005). In other studies Rashid et al. (2000), Hafeez et al. (2002) and Anjum et al. (2007) observed that bacterial inoculum significantly increased seed cotton yield, and plant height over their respective controls. Increase in the biomass could be due to better nutrition of the plant under saline conditions. Erdogan and Benlioglu (2010) reported that seed treatment of cotton plants with *Pseudomonas* spp. and *Serratia plymuthica* improved growth parameters of cotton in field experiments.

The positive effects of PGPR on plant growth are always correlated with remarkable changes in root morphology, namely increased lateral root length and root hair number and length (Lugtenberg and Kamilova, 2009). Increased number of lateral roots may result in better absorption of water and nutrients from the soil (Hoflich and Kuhn 1996). Similar results we observed in our previous work (Egamberdiyeva, 2007) in which in nutrient-poor arid soil, bacterial strains significantly ($P < 0.05$) enhanced early plant growth of maize, and the inoculation could compensate for nutrient deficiency and improve plants' development. In addition, several physiological, and biochemical changes in plant due to inoculation of rhizobacteria have been suggested to help in alleviating the salt stress (Han and Lee, 2005; Barriuso et al., 2008). In the present work we tried to evaluate on which mechanisms the observed plant growth stimulation in saline soil can be based. Some mechanisms such as production of phytohormones (Frankenberger and Arshad, 1995; Spaepen et al., 2009) and ACC deaminase were demonstrated.

It is known that unfavourable environmental factors such as salinity and drought cause sharp changes in the balance of phytohormones associated with a decline in the level of growth-activating hormones such as IAA (Sakhabutdinova et al., 2003). The exogenous application of plant growth regulators (PGRs) e.g. gibberellins (Afzal et al., 2005), auxins (Khan et al., 2004), cytokinins (Gul et al., 2000) produced some benefit in alleviating the adverse effects of salt stress and they also improve germination, growth, fruit setting, fresh vegetable and seed yields and yield quality (Saimbhi, 1993). The root colonising IAA-producing bacteria may supply additional phytohormones to the plant, and stimulate root growth (Afzal et al., 2005). In our study two effective strains *P. putida* BIST and *B. subtilis* SUBTIN were able to produce IAA in saline condition. Plant stress can be reduced by 1-aminocyclopropane-carboxylate (ACC) deaminase-producing bacteria (Glick et al., 1997). Our strain *P. putida* BIST showed ACC deaminase activity. Yue *et al.* (2007) also demonstrated that *Klebsiella oxytoca* strain Rs-5, which had ACC deaminase activity, could relieve salt stress and promote cotton seedling growth in saline soil (Glick et al., 2007). The bacteria contain the enzyme ACC deaminase and colonize to the seed coat of a developing seedling, may decrease the ethylene level so as it does not become too high to limit where root growth (Hontzeas et al., 2004). These bacteria may enhance the survival of some seedlings, especially during the first few days after the seeds are planted. On the basis of results obtained for this study, it may be concluded that rhizobacteria are potential option for improvement of cotton growth and development in salinated arid soil.

References

Adesemoye ,A.O. & Kloepper, J.W. (2009). Plant-microbes interactions in enhanced fertilizer use efficiency. *Appl. Microbiol. Biotechnol.* 85:1–12.

Afzal, I., Basra, Sh., Iqbal, A. (2005). The effect of seed soaking with plant growth regulators on seedling vigor of wheat under salinity stress. *J. Stress Physiol. Biochem.* 1(1):6-14.

Anjum, M.A., Saijad, M.R., Akhtar, N., Qureshi, M.A., Iqbal, A., Jami, A.R, Hasan, M. (2007). Response of cotton to plant growth promoting rhizobacteria (PGPR) inoculation under different levels of nitrogen. *J. Agri. Res.* 45 (2):135-143.

Arora, N.K., Khare, E., Oh, J.H., Kang, S.C. & Maheshwari, D.K. (2008). Diverse mechanisms adopted by fluorescent *Pseudomonas* PGC2 during the inhibition of *Rhizoctonia solani* and *Phytopthora capsisi. World. J. Microbiol. Biotechnol.* 24:581-585.

Bano, N. & Musarrat, J. (2003). Characterization of a new Pseudomonas aeruginosa strain NJ-15 as a potential biocontrol agent. *Curr. Microbiol.* 46:324-328.

Barriuso, J., Ramos, S.B. & Gutierrez Manero, F.J. (2008). Protection against pathogen and salt stress by four plant growth-promoting rhizobacteria isolated from *Pinus* sp. On *Arabidopsis thaliana*. *Phytopathol.* 98:666–672.

Burdman, S., Jurkevitch, E. & Okon, Y. (2000). *Recent advances in the use of plant growth promoting rhizobacteria (PGPR) in agriculture, in: Microbial Interactions in Agriculture and Forestry*. N. S. Subba Rao and Y. R. Dommergues, eds., Science Publishers, Enfield, USA, Vol II. 229-250.

Chakraborty, U., Chakraborty, B.N., Basnet, M. & Chakraborty, A.P. (2009). Evaluation of *Ochrobactrum anthropi* TRS-2 and its talc based formulation for enhancement of growth of tea plants and management of brown root rot disease. *J. Appl. Microbiol.* 107(2):625-634

DIN ISO 15178. (2001). *Bodenbeschaffenheit - Bestimmung des Gesamt-Schwefels durch trockene Verbrennung (Elementaranalyse)*. Beuth, Berlin Wien Zürich.

DIN ISO 38414-S. (1983). *Deutsche Einheitsverfahren zur Wasser, Abwasser und Schlammuntersuchung: Schlamm und Sdediment*. Beuth, Berlin Wien Zürich.

Duffy, B., Keel, C., Defago, G. (2004). Potential role of pathogen signaling in multitrophic plant-microbe interactions involved in disease protection. *Appl. Environ. Microbiol.* 70(3):1836–1842.

El-Sayed M. & Embaby. (2007). Control of damping off and or sore shin in cotton and white mould in cowpea plant disease(s) by using a bio-fungicide *Coniothyrium minitans* campbell. *Res. J. Agric. Biolog. Sci.* 3(4):267-273.

Egamberdieva, D., Kucharova, Z., Davranov, K., Berg, G., Makarova, N., Azarova, T., Chebotar, V., Tikhonovich, I., Kamilova, F., Validov, Sh. & Lugtenberg, B. (2010). Bacteria able to control foot and root rot and to promote growth of cucumber in salinated soils. *Biol. Fertil. Soil.* 47:197-205.

Egamberdieva, D. (2008). Plant growth promoting properties of rhizobacteria isolated from wheat and pea grown in loamy sand soil. *Turk. J. Biol.* 32(1):9-15.

.Egamberdieva, D., Kamilova, F., Validov, S., Gafurova, L., Kucharova, Z. & Lugtenberg, B. (2008). High incidence of plant growth-stimulating bacteria associated with the rhizosphere of wheat grown on salinated soil in Uzbekistan. *Env. Microbiol.* 10:1-9.

Egamberdiyeva, D. &Islam, K.R. (2008). Salt tolerant rhizobacteria: Plant growth promoting traits and physiological characterization within ecologically stressed environment. In: Ahmad I, Pichtel J, Hayat Sh (eds) *Plant-Bacteria Interactions: strategies and techniques to promote plant growth*, Wiley-VCH Verlag GmbH & Co. Weinheim. Pp. 257-281.

Egamberdiyeva, D., Gafurova, L. & Islam, K.R. (2007). Salinity Effects on Irrigated Soil Chemical and Biological Properties in the Syr Darya Basin of Uzbekistan. In: Lal, R., Sulaimanov, M., Stewart, B., Hansen, D. and Doraiswamy, P. (Eds.). *Climate Change and Terrestrial C Sequestration in Central Asia,* Taylor-Francis, New York.

Egamberdieva, D. & Kucharova, Z. (2009). Selection for root colonising bacteria stimulating wheat growth in saline soils. *Biol. Fertil. Soil.* 45:561-573. El-Sayed, M. & Embaby. (2007). Control of damping off and/or sore shin in cotton and white mould in cowpea plant disease(s) by using a bio-fungicide Coniothyrium minitans Campbell. *Res. J. Agri. Biolog. Sci.* 3(4): 267-273.

Erdogan, O. & Benlioglu, K. (2010). Biological control of Verticillium wilt on cotton by the use of fluorescent *Pseudomonas spp.* under field conditions. *Biolog. Cont*. 53, 39–45.

Frankenberger, J.W.T. & Arshad, M. (1995). Microbial synthesis of auxins. In: Frankenberger WT, Arshad M, (eds.) *Phytohormones in soils*. New York: Marcel Dekker, Inc. 35–71.

Glick, B. R., Cheng, Z., Czarny, J. & Duan, J. (2007). Promotion of plant growth by ACC deaminase-producing soil bacteria. *Eur. J. Plant Pathol.* 119:329-339.

Glick, B.R., Todorovic, B., Czarny, J., Cheng, Z., Duan, J. & McConkey, B. (2007). Promotion of plant growth by bacterial ACC deaminase. *Crit. Rev. Plant. Sci.* 26:227-242.

Glick, B.R., Liu, C., Ghosh, S. & Dumbrof, E.B. (1997). Early development of canola seedlings in the presence of the plant growth-promoting rhizobacterium *Pseudomonas putida* GR12-2. *Soil. Biol. Biochem.* 29:1233–1239.

Gul, B., Khan, M.A. & Weber, D.J. (2000). Alleviation salinity and dark-enforced dormancy in *Allenrolfea occidentalis* seeds under various thermoperiods. *Aust. J. Bot*. 48:745–752.

Hafeez, F.Y., Safdar, M.E., Ali, Z., Rasool, G., Shakeel, M., Malik, K.A. (2002). Isolation and characterization of rhizobial and PGPR strains from rhizosphere of cotton and their effect on growth of cotton. *Abstracts of 9th International Congress of Soil Sciences*, March 18-20, NIAB. 34.

Han, H.S. & Lee, K. D. (2005). Plant growth promoting rhizobacteria effect on antioxidant status, photosynthesis, mineral uptake and growth of lettuce under soil salinity. *Res. J. Agri. Biol. Sci.* 1:210–215.

Hassanin, S.M., El-Mehalawy, A.A., Hassanin, N.M & Zaki, S.A. (2007). Induction of resistance and biocontrol of rhizoctonia in cotton damping-off disease by rhizosphere bacteria and actinomycetes. *The Internet J. Microbiol.* 3 (2).

Hontzeas, N., Saleh, S. and Glick, B.R. (2004) Changes in gene expression in canola roots induced by ACC-deaminase-containing plant-growth-promoting bacteria. *Mol. Plant-Microbe Interact.* 12:951–959.

Höflich, G. & Kühn, G. (1996). Förderung das Wachstums und der Nährstoffaufnahme bei kurziferen Öl- und Zwischenfruhten durch inokulierte Rhizospherenmikroorganismen. *Zeit. Pflanzen. Bod.* 159:575-578.

Khan, M.A., Gul, B. & Darrell, J. (2004). Weber Action of plant growth regulators and salinity on seed germination of *Ceratoides lanata. Can. J. Bot.* 82:37–42.

Leeman, J.A., Pelt, V., den Puden, V.M., Heinsbroek, M., Bakker, P.A.H. & Schippers, B. (1995). lnduction of systemic resistance by Pseudomonas fluorescens in radish cultivars differing in susceptibility to Fusarium wilt, using a novel bioassay. *Europ. J. Plant. Pathol.* 101:655-664.

Lugtenberg, B.J.J., Dekkers, L. & Bloemberg, G.V. (2001). Molecular determinants of rhizosphere colonization by Pseudomanas. *Annu. Rev. Phytopath.* 39:461-490.

Lugtenberg, B.J.J. & Kamilova, F. (2004). Rhizosphere management: microbial manipulation for biocontrol. In: Goodman RM (Eds) *Encyclopedia of plant and crop science*, New York, Marcel Dekker. pp. 1098-1101.

Lugtenberg B.J.J. & Kamilova, F. (2009). Plant growth-promoting rhizobacteria. *Annual. Rev. Microbiol.* 63:541-556.

Lugtenberg, B.J.J., Kravchenko, L.V. & Simons, M (1999) Tomato seed and root exudate sugars: composition, utilization by Pseudomonas biocontrol strains and role in rhizosphere colonization. *Environ. Microbiol.* 1:439-446.

Maheshwari, D.K., Kumar, S., Kumar, B. & Pandey, P. (2011). Co-inoculation of urea and DAP tolerant *Sinorhizobium meliloti* and *Pseudomonas aeruginosa* as integrated approach for growth enhancement of *Brassica juncea*. *Ind. J. Microbiol.* 50(4):425-437.

Mayak, S., Tirosh, T. & Glick, B.R. (2004). Plant growth-promoting bacteria that confer resistance to water stress in tomatoes and peppers. *Plant. Sci.* 166:525–530.

Patten, C.L. & Glick, B.R. (2002). Role of *Pseudomonas putida* indoleacetic acid in development of the host plant root system. *Appl. Environ. Microbiol.* 68:3795-3801.

Rashid, A., Siddique, G & Sindhu, M.S. (2000). *Azotobacter* and *Azospirillum* inoculation effect on cotton yield. *Abstracts of 8th International Congress of Soil Sciences,* November, NARC, Islamabad. 28.

Requena, B.N., Jimenez, I., Toro, M. & Barea. J.M. (1997). Interactions between plant-growth-promoting rhizobacteria (PGPR), arbuscular mycorrhizal fungi and *Rhizobium* spp. in the rhizosphere of *Anthyllis cytiisoides*, a model legume for revegetation in mediterranean semi-arid ecosystems. *New. Phytol.* 136:667-677.

Sakhabutdinova, A.R., Fatkhutdinova, D.R., Bezrukova, M.V. & Shakirova, F.M. (2003). Salicylic acid prevents the damaging action of stress factors on wheat plants. Bulg. *J. Plant. Physiol.* 314-319.

Sheraliev, A., Sattarova, R. & Rahimov, U. (2008). *Agri. Phytopathol. Tashkent,* Uzbekistan, 250.

Spaepen, S., Vanderleyden, J. & Okon, Y. (2009). Plant Growth-promoting actions of rhizobacteria. In van Loon LC, Ed Kader JK, Delseny M (Eds). *Advan. Botan. Res.* 51:283-320.

Shirokova, Y., Forkutsa, I. & Sharafutdinova, N. (2000). Use of electrical conductivity instead of soluble salts for soil salinity monitoring in Central Asia. *Irrig. Drain. Syst.* 14:199-205.

Werner, J.E. & Finkelstein, R.R. (1995). Arabidopsis mutants with reduced response to NaCl and osmotic stress. *Physiol. Plantar.* 93:659-666.

Woodward, A.W. & Bartel, B. (2005). Auxin: regulation, action and interaction. *Annals. Bot.* 95:707–735.

Yue, H., Mo, W., Li, C., Zheng, Y. & Li, H. (2007). The salt stress relief and growth promotion effect of Rs-5 on cotton. *Plant*. Soil. 297:139-145.

Zahir, Z.A., Munir, A., Asghar, H.N., Shaharoona, B. & Arshad, M. (2008). Effectiveness of rhizobacteria containing ACC deaminase for growth promotion of peas (*Pisum sativum*) under drought conditions. *J. Microbiol. Biotechnol.* 18:958-963.

INDEX

A

B

C

D

E

F

G

H

I

J

K

L

M

N

O

P

Q

R

S

T

U